Quasi-Ordinary Power Series and Their Zeta Functions

MEMOIRS
of the
American Mathematical Society

Number 841

Quasi-Ordinary Power Series and Their Zeta Functions

Enrique Artal Bartolo
Pierrette Cassou-Noguès
Ignacio Luengo
Alejandro Melle Hernández

November 2005 • Volume 178 • Number 841 (end of volume) • ISSN 0065-9266

American Mathematical Society
Providence, Rhode Island

2000 *Mathematics Subject Classification.*
Primary 14B05, 14E15, 32S50.

Library of Congress Cataloging-in-Publication Data

Quasi-ordinary power series and their zeta functions / Enrique Artal Bartolo ... [et al.].
p. cm. — (Memoirs of the American Mathematical Society, ISSN 0065-9266 ; no. 841)
"Volume 178, number 841 (end of volume)."
Includes bibliographical references.
ISBN 0-8218-3876-8 (alk. paper)
1. Singularities (Mathematics). 2. Power series. 3. Functions, Zeta. I. Artal-Bartolo, Enrique, 1963– II. Series.

QA3.A57 no. 841
[QA614.58]
510 s—dc22
[516.3′5] 2005050801

Memoirs of the American Mathematical Society

This journal is devoted entirely to research in pure and applied mathematics.

Subscription information. The 2005 subscription begins with volume 173 and consists of six mailings, each containing one or more numbers. Subscription prices for 2005 are $606 list, $485 institutional member. A late charge of 10% of the subscription price will be imposed on orders received from nonmembers after January 1 of the subscription year. Subscribers outside the United States and India must pay a postage surcharge of $31; subscribers in India must pay a postage surcharge of $43. Expedited delivery to destinations in North America $35; elsewhere $130. Each number may be ordered separately; *please specify number* when ordering an individual number. For prices and titles of recently released numbers, see the New Publications sections of the *Notices of the American Mathematical Society.*

Back number information. For back issues see the *AMS Catalog of Publications.*

Subscriptions and orders should be addressed to the American Mathematical Society, P. O. Box 845904, Boston, MA 02284-5904, USA. *All orders must be accompanied by payment.* Other correspondence should be addressed to 201 Charles Street, Providence, RI 02904-2294, USA.

Memoirs of the American Mathematical Society is published bimonthly (each volume consisting usually of more than one number) by the American Mathematical Society at 201 Charles Street, Providence, RI 02904-2294, USA. Periodicals postage paid at Providence, RI. Postmaster: Send address changes to Memoirs, American Mathematical Society, 201 Charles Street, Providence, RI 02904-2294, USA.

This publication is indexed in *Science Citation Index®*, *SciSearch®*, *Research Alert®*, *CompuMath Citation Index®*, *Current Contents®/Physical, Chemical & Earth Sciences.*
Printed in the United States of America.

♾ The paper used in this book is acid-free and falls within the guidelines
established to ensure permanence and durability.
Visit the AMS home page at `http://www.ams.org/`

10 9 8 7 6 5 4 3 2 1 10 09 08 07 06 05

Contents

Introduction 1

Chapter 1. Motivic integration 7
1. Grothendieck ring of varieties 7
2. The arc space of a variety 7
3. Local Denef-Loeser motivic zeta function 8

Chapter 2. Generating functions and Newton polyhedra 11
1. Generating functions for integer points in rational polyhedra 11
2. Motivic zeta function and Newton polyhedra 13

Chapter 3. Quasi-ordinary power series 21
1. Characteristic exponents 22
2. Newton polyhedron and good coordinates 23
3. Dual decomposition 26
4. Newton map associated with a Newton component 29
5. Transversal sections of a quasi-ordinary power series 37

Chapter 4. Denef-Loeser motivic zeta function under the Newton maps 41
1. Vertices of the dual decomposition 46
2. Edges of the Newton polytope 49
3. Zeta functions along strata 59

Chapter 5. Consequences of the main theorems 63
1. Essential variables 63
2. Curve case 63
3. The topological zeta function 64
4. A special candidate pole 65

Chapter 6. Monodromy conjecture for quasi-ordinary power series 77
1. Monodromy conjecture for curves 77
2. Monodromy conjecture: general case 79
3. Monodromy conjecture for the Igusa zeta-function 80

Bibliography 83

Abstract

The main objective of this paper is to prove the monodromy conjecture for the local Igusa zeta function of a quasi-ordinary polynomial of arbitrary dimension defined over a number field. In order to do it, we compute the local Denef-Loeser motivic zeta function $Z_{\rm DL}(h,T)$ of a quasi-ordinary power series h of arbitrary dimension over an algebraically closed field of characteristic zero from its characteristic exponents without using embedded resolution of singularities. This allows us to effectively represent $Z_{\rm DL}(h,T) = P(T)/Q(T)$ such that almost all the candidate poles given by $Q(T)$ are poles. Anyway, these candidate poles give eigenvalues of the monodromy action on the complex $R\psi_h$ of nearby cycles on $h^{-1}(0)$. In particular we prove in this case the monodromy conjecture made by Denef-Loeser for the local motivic zeta function and the local topological zeta function. As a consequence, if h is a quasi-ordinary polynomial defined over a number field we prove the Igusa monodromy conjecture for its local Igusa zeta function.

Received by the editor August 1, 2003.

2000 *Mathematics Subject Classification.* 14B05,14E15,32S50.

Key words and phrases. Motivic, topological and Igusa zeta functions, monodromy, quasi-ordinary singularities.

First author is partially supported by BFM2001-1488-C02-02; the last two authors are partially supported by BFM2001-1488-C02-01.

Introduction

Let h be a polynomial in $\mathbb{Z}[x_1, \ldots, x_d]$ and fix $p \in \mathbb{Z}$ a prime number. In order to study the number N_k of solutions of the congruence $h \equiv 0 \bmod p^k$, classically one associates with h the Poincaré series

$$P(T) = \sum_{k=0}^{\infty} N_k T^k.$$

J. Igusa proved in [**25, 26**] that $P(T)$ is a rational function on T by relating it with the following p-adic integral

$$I(h,s) := \int_{\mathbb{Z}_p^d} |h(x)|^s |dx|, \tag{0.1}$$

for $s \in \mathbb{C}$, $\mathrm{Re}(s) > 0$, where $|dx|$ denotes the Haar measure on $\mathbb{Q}_p^d$ normalized in such a way that $\mathbb{Z}_p^d$ is of volume 1.

Igusa proved the identity $P(p^{-d-s}) = \frac{1-p^{-s}I(h,s)}{1-p^{-s}}$ and he used an embedded resolution of $h^{-1}(0)$ to show that $I(h,s)$ is a rational function of p^{-s} (see [**9**] for a proof without resolution of singularities). In fact each exceptional divisor of an embedded resolution gives a candidate pole of $I(h,s)$ but many of them are not actually poles.

Igusa conjectured that poles of $I(h,s)$ are related to eigenvalues of the complex monodromy at some point of $h^{-1}(0)$, see [**10**]. More precisely he raised the *conjecture*:

Igusa Monodromy Conjecture. *If $h \in F[x_1, \ldots, x_d] \setminus F$, for some number field $F \subset \mathbb{C}$, then for almost all p-adic completions K of F, if s_0 is a pole of $I(h,K,s)$, then* $\exp(2i\pi\Re(s_0))$ *is an eigenvalue of the local monodromy of h at some complex point of $h^{-1}(0)$.*

Since then, some partial results have been obtained, see the Bourbaki Seminar talk by J. Denef [**10**] for a survey of these results until 1991, and for instance the papers [**33, 34, 44, 45, 40, 2**] for more recent results. We recall two cases where Igusa monodromy conjecture has been proved. In fact in both cases, the following stronger version of the Igusa monodromy conjecture was proved:

For almost all p-adic completions K of F, if s_0 is a pole of $I(h,K,s)$, then $\Re(s_0)$ is a root of the Bernstein polynomial $b_h(s)$ of h.

Firstly, F. Loeser in [**33**] gave a proof for reduced polynomials in two variables. One other interesting case is the case of polynomials (in arbitrary dimension) non-degenerate with respect to their Newton polyhedron and satisfying some "resonance" conditions. This result was also proved by F. Loeser in [**34**]. For plane curves, the strong candidate poles come from the rupture components in an embedded resolution of $h^{-1}(0)$, see [**45, 46**]. For non-degenerate polynomials J. Denef gave a set of strong candidate poles which comes from faces of codimension 1 in the Newton polyhedron of h, see [**11**].

In this paper we will prove the Igusa monodromy conjecture for the local Igusa zeta function $I_0(h, K, s)$, which is the local version of $I(h, K, s)$, for quasi-ordinary polynomials in arbitrary dimension. As we will see quasi-ordinary polynomials behave in many aspects as plane curves. In our proof we use in an essential way motivic integration on the space of arcs on an algebraic variety.

Motivic measures take values in a completion of the Grothendieck ring of algebraic varieties. Let $K_0(\mathrm{Var}_k)$ be the Grothendieck ring of algebraic varieties over k. Let $\mathbb{L} = [\mathbb{A}_k^1]$ denote the class in $K_0(\mathrm{Var}_k)$ of the affine line. The naive motivic ring $\mathcal{M}_k$ of algebraic varieties over k is the polynomial ring $\mathcal{M}_k = K_0(\mathrm{Var}_k)[\mathbb{L}^{-1}]$.

Let X be a non singular irreducible complex algebraic variety of pure dimension d. For any $n \in \mathbb{N}$, let $\mathcal{L}_n(X)$ denote the space of arcs modulo t^{n+1} on X; it has a structure of complex variety. The arc space $\mathcal{L}(X)$ of X is the projective limit of the algebraic varieties $\mathcal{L}_n(X)$. For any $n \in \mathbb{N}$, let $\pi_n : \mathcal{L}(X) \to \mathcal{L}_n(X)$ be the natural projection. For any arc $\varphi \in \mathcal{L}(X)$, the origin of the arc is $\pi_0(\varphi)$. For each closed point $x \in X$, let $\mathcal{L}_x(X)$ (resp. $\mathcal{L}_{n,x}(X)$) be the set of arcs on X (resp. truncated arcs) with origin at x.

Consider $X = \mathbb{A}^d$ the d-dimensional complex affine space and $x = \mathbf{0}$ its origin. Let $h \in \mathbb{C}[x_1, \ldots, x_d]$ be a complex polynomial, with $h(\mathbf{0}) = 0$. Set $V_n := \{\varphi \in \mathcal{L}_x(X) : \mathrm{ord}(h \circ \varphi) = n\}$. Denef and Loeser in [**17**] defined the *naive motivic zeta function* of h by

$$Z_{\mathrm{naive}}(h, T) := \sum_{n \geq 1} [\pi_n(V_n)] \mathbb{L}^{-nd} T^n \in \mathcal{M}_k[[T]].$$

In fact we will consider the local Denef-Loeser motivic zeta function which is $Z_{DL}(h, T) = \mathbb{L}^{-d} Z_{\mathrm{naive}}(h, T)$. They showed in [**15**], using embedded resolution of singularities, that $Z_{DL}(h, T)$ is a rational function. It belongs to the subring $\mathcal{N}$ of the ring $\widehat{\mathcal{M}}_k[[T]]$ which is generated by the image in $\widehat{\mathcal{M}}_k[[T]]$ of $\mathcal{M}_k[T]$ and $(1 - \mathbb{L}^{-a} T^b)^{-1}$, $a, b \in \mathbb{N}, b > 0$. The monodromy conjecture in this case states, see [**15**, section 2.4]:

MOTIVIC MONODROMY CONJECTURE. *There is a set $S = \{(a, b) : a, b \in \mathbb{N}, b > 0\}$ such that $Z_{DL}(h, T) \in \mathcal{M}_k[T][(1 - \mathbb{L}^{-a} T^b)^{-1}]_{(a,b) \in S}$ and if $q = a/b$, $(a, b) \in S$, then $\exp(2i\pi q)$ is an eigenvalue of the local complex algebraic monodromy around zero at some $P \in h^{-1}(0)$.*

It turns out that $Z_{DL}(h, T)$ is the right function to study several monodromy conjectures with because it specializes to the local Igusa zeta function and to the local topological zeta function:

- If h is a non-constant polynomial defined over a number field F then it follows from [**15**], see also [**17**], that for almost all finite places of F, the real parts $Ns + \nu$ of poles of $I_0(h, K, s)$ come from factors $(1 - \mathbb{L}^{-\nu} T^N)$ in the denominator of $Z_{\mathrm{naive}}(h, T)$.
- The local topological zeta function $Z_{\mathrm{top},0}(h, s)$ of a complex polynomial h was introduced by Denef and Loeser in [**14**] as a kind of limit of the local Igusa zeta function. Later on, they obtained $Z_{\mathrm{top},0}(h, s)$ from $Z_{DL}(h, T)$ by the following procedure, cf. [**15**, § 2.3]. First substitute $\mathbb{L}^{-s}$ for T in $Z_{DL}(h, T)$, then expand

$\mathbb{L}^{-s}$ and $(\mathbb{L}-1)(1-\mathbb{L}^{-\nu+Ns})^{-1}$ into series in $\mathbb{L}-1$. Finally take the usual Euler characteristic χ_{top} of the result (this works because $\chi_{\text{top}}(\mathbb{L})=1$). Then the local topological zeta function is the following rational function, see [**17, 15**]:

$$Z_{\text{top},0}(h,s) := \chi_{\text{top}}(Z_{DL}(h,\mathbb{L}^{-s})).$$

The local topological zeta function is a subtle analytic invariant (but not topological, see [**3**]) of the local singularity. The poles of $Z_{\text{top},0}(h,s)$ induce poles of $Z_{DL}(h,T)$. The other way around is not true, see Example 2.10. In this case Denef and Loeser conjectured in [**14**]:

TOPOLOGICAL MONODROMY CONJECTURE. *If $s_0 = -\frac{\nu}{N}$ is a pole of $Z_{\text{top},0}(h,s)$ then $\exp(2i\pi(-s_0))$ is an eigenvalue of the monodromy action at some point of $h^{-1}(0)$.*

One result of this paper is that if h defines a quasi-ordinary singularity of hypersurface of arbitrary dimension then the monodromy conjecture for $Z_{DL}(h,T)$ is true, see Corollary 6.5. This fact implies the Igusa monodromy conjecture for $I_0(h,K,s)$ and the Denef-Loeser monodromy conjecture for $Z_{\text{top},0}(h,s)$. In fact we prove a stronger result.

A germ of a complex analytic variety $(V,0)$ is called a *quasi-ordinary singularity* if there is a finite morphism (proper with finite fiber map) of analytic germs $\pi : (V,0) \to (\mathbb{C}^d,0)$ whose discriminant locus is contained in $x_1x_2\ldots x_d=0$, for some local coordinates $(x_1,\ldots,x_d)$. We are only interested in quasi-ordinary hypersurface singularities $(V,0) \subset (\mathbb{C}^{d+1},0)$.

A convergent power series $h \in \mathbb{C}\{\mathbf{x},z\}$, $h(0)=0$, defines a *quasi-ordinary singularity* at zero if the germ of its zero locus $(V,0) \subset (\mathbb{C}^{d+1},0)$ is a *quasi-ordinary singularity*. These singularities behave in many aspects as singularities of plane curves mainly because, after Jung-Abhyankar theorem, they admit fractional power series parameterizations and a finite set of *characteristic exponents*, see [**28**] and [**1**] for an algebraic proof.

Quasi-ordinary singularities appears early in the problem of resolution of singularities of surfaces through the work of Jung ([**28**], [**49**]) as follows. Let $p : S \subset \mathbb{C}^3 \to \mathbb{C}^2$ be a finite linear projection of the algebraic surface S with discriminant locus $\Delta \subset \mathbb{C}^2$. Take an embedded resolution $\pi : \widetilde{\mathbb{C}^2} \to \mathbb{C}^2$ of the curve Δ where the exceptional locus $\mathcal{D} := \pi^{-1}(\Delta)$ is a normal crossing divisor. The pull-back $\tilde{p} : \tilde{S} \to \widetilde{\mathbb{C}^2}$ of p is a finite map whose discriminant is contained in $\mathcal{D}$. Thus the singularities of $\tilde{S}$ are all quasi-ordinary and can be resolved easily. Namely, if $n : Z \to \tilde{S}$ is the normalization of $\tilde{S}$ then Z is smooth over simple points of $(\mathcal{D})_{\text{red}}$ and above the remaining points Z have cyclic quotient singularities, see for instance Laufer's book [**30**] for a detailed explanation and several examples.

In fact, for surfaces the resolution can be done without normalization which allows one to get also an embedded resolution, as explained by O. Zariski in the nice papers [**50, 51**], in which he introduced the name *quasi-ordinary singularities* and defined the characteristic exponents. J. Lipman in his Harvard thesis [**31**] studied in more details the resolution of quasi-ordinary singularities of surfaces proving that the characteristic exponents, suitable normalized, are independent of the chosen coordinates. In recent years the interest on quasi-ordinary singularities in any dimension has been renewed thanks to many interested results and many interested open problems (see for instance the paper by L.J. McEwan and A. Némethi in [**37**]). We cite some of them and of course the list is far to be complete.

J. Lipman and Y. Gau proved that in the irreducible case these exponents determine the embedded topology of $(V, 0)$, [**19, 32**]. Popescu-Pampu [**39**] defined the semigroup of a quasi-ordinary singularity proving its invariance, see also [**21**]. P.D. González Pérez [**20, 22**] has constructed an embedded resolution of a quasi-ordinary singularity which depends only on its characteristic exponents. This last result and a different embedded resolution method for quasi-ordinary singularites constructed by O. Villamayor [**48**], allows in principle to compute the Denef-Loeser zeta function for quasi-ordinary singularites. Here we will propose a different approach based on Newton maps.

In [**36**], I. Luengo showed that the usual Newton-Puiseux method for curves can be used in arbitrary dimension to find the roots with fractional exponents of a quasi-ordinary polynomial by means of *Newton maps* (see Chapter 4). The key point to find the roots is that after a Newton map we get a new quasi-ordinary polynomial with fewer characteristic exponents and one proceeds by finite induction. Instead of using the embedded resolution of quasi-ordinary singularities to compute $Z_{DL}(h, T)$, we use the above procedure given by Newton maps. This method allows us to decompose $Z_{DL}(h, T)$ as a sum of two rational motivic zeta functions:

$$Z_{DL}(h, T) = Z^A_{DL}(h, T) + Z^B_{DL}(h, T).$$

The A-part corresponds to arcs $\varphi \in \mathcal{L}_0(X)$ such that the t-order $\mathrm{ord}_t(h \circ \varphi)$ can be computed from the (degenerate or not) Newton polytope of h. The computation of $Z^A_{DL}(h, T)$ from the Newton polytope follows ideas of J. Denef and K. Hoornaert in the p-adic case, see [**12**]. In fact we will show an interesting description of $Z^A_{DL}(h, T)$ in terms of generating functions of some rational polyhedra obtained from the Newton polytope of h, see Chapters 1 and 2. As an application, a formula for $Z_{DL}(h, T)$ for any germ of complex analytic function h with non-degenerate Newton polytope is given, see theorem 2.4. G. Guibert has recently obtained a similar formula, see [**24**].

In order to compute the measure of the arcs $\varphi \in \mathcal{L}_0(X)$ in the B-part we pull back these arcs under Newton maps. In particular $Z^B_{DL}(h, T)$ is the sum of some motivic zeta functions depending on the pull-back of h under all its Newton maps. There is one major technical problem here. In dimension higher than 1, usually there exist some arcs which cannot be lifted under usual Newton maps. To solve this problem we need to consider Newton maps with coefficients in $\mathbb{C}\{t\}$. Thus our results really deal with quasi-ordinary power series with coefficients in $\mathbb{C}\{t\}$ instead of $\mathbb{C}$. Now Newton maps are $\mathbb{C}\{t\}$-morphisms in the terminology introduced by Denef and Loeser in [**18**] and we can apply the change of variables formula. The differential form plays a role here but throughout this introduction we omit it. In fact, we perform all these computations for an algebraically closed field k of characteristic zero.

Essentially in $Z^B_{DL}(h, T)$ we get quasi-ordinary singularities with fewer characteristic exponents and we can apply recursively this formula. In this way, for $Z_{\mathrm{top},0}(h, s)$ a very effective and closed recursive formula is given only depending on the tree of characteristic exponents, cf. Theorem 5.3. For $Z_{DL}(h, T)$ the formula, also recursive, is enough to give a short list of candidate poles. Each characteristic exponent is a rational d-tuple. Each non-zero coordinate of each characteristic exponent will give a candidate pole of $Z_{DL}(h, T)$. In section 6 some of them are excluded to get a smallest set $SCP(h)$ of strong candidate poles. Our main result is

$$(0.2)\qquad Z_{DL}(h,T) \in \mathbb{Z}[\mathbb{L},\mathbb{L}^{-1},(1-\mathbb{L}^{-\nu}T^{N})^{-1}][T]_{(N,\nu)\in SCP(h)}.$$

The proof of the above result gives some extra information. Namely all elements in $SCP(h)$, but one concrete case $(*)$ when $d=2$ (cf. Proposition 5.10), appear also as strong candidate poles of a transversal section of h at some point of the singular locus of $h^{-1}(0)$. Transversal sections at generic points are also quasi-ordinary singularities, this fact allows us to prove the motivic monodromy conjecture by induction on the dimension. For curves, this method gives a closed formula for $Z_{DL}(h,T)$, cf. (5.1), and from this formula a new proof of the monodromy conjecture follows directly. In the $(*)$-case, there is only one strong candidate pole not appearing in the transversal sections. We use the formula, proved by L.J. McEwan and A. Némethi in [**38**] (in the irreducible case) and by P.D. González Pérez, L.J. McEwan and A. Némethi in [**23**] (in general), for the zeta function of the monodromy at the origin of quasi-ordinary singularities to show that this strong candidate pole also gives an eigenvalue of the monodromy of h at the origin.

The computation of $Z_{\mathrm{top},0}(h,s)$ for quasi-ordinary singularities gives a very effective way to compute the poles of $Z_{\mathrm{top},0}(h,s)$ for a general surface singularity. To shorten this work we will provide the details in a forthcoming paper. The basic idea is to use the well-known *Jung-method* as follows.

Let $p:(S,0)\subset(\mathbb{C}^3,0)\to(\mathbb{C}^2,0)$ be a finite morphism of a (hyper)surface singularity $(S,0)$, defined by h, with discriminant locus $(\Delta,0)\subset(\mathbb{C}^2,0)$. Take an embedded resolution $\pi:(\widetilde{\mathbb{C}^2},\mathcal{D})\to(\mathbb{C}^2,0)$ of the germ of curve $(\Delta,0)$ where the exceptional locus $\mathcal{D}=\pi^{-1}(0)$ is a normal crossing divisor. The pull-back $\tilde{p}:(\tilde{S},\mathcal{E})\to(\widetilde{\mathbb{C}^2},\mathcal{D})$ is a finite map whose discriminant is contained in the pull-back of $(\Delta,0)$. Thus the singularities of $(\tilde{S},\mathcal{E})$ are all quasi-ordinary. We have proved that the poles of $Z_{\mathrm{top},0}(h,s)$ are contained in the set of poles of either the transversal sections (now curves) at the rupture components of the resolution or components of the singular locus of $(S,0)$. This gives a short list of strong candidate poles which are in general poles of $Z_{\mathrm{top},0}(h,s)$. The next step will be to prove that they induce eigenvalues of the monodromy.

CONVENTIONS. Throughout this paper we denote by $\mathbb{N}$ the set of the non-negative integers, $\mathbb{P}$ the set of positive integers, $\mathbb{R}_+=\{x\in\mathbb{R}: x\geq 0\}$ and $\mathbb{R}_{>0}=\{x\in\mathbb{R}: x>0\}$. To shorten the notation we will use bold symbols for d-tuples, for instance $\mathbf{x}=(x_1,\dots,x_d)$.

In this paper we work over a field k of characteristic zero. A variety over k will mean a reduced separated scheme of finite type over the field k, $\mathbf{G}^d_{m,k}:=\operatorname{Spec}k[x_1,\dots,x_d,x_1^{-1},\dots,x_d^{-1}]$ denote the d-dimensional torus over k and $\mathbb{A}^d_k:=\operatorname{Spec}k[x_1,\dots,x_d]$ the d-dimensional affine space over k.

ACKNOWLEDGMENT. The authors thank the referees for many interesting comments and remarks about this paper which have really improved the quality of the text. The authors also thank H. Cobo Pablos, J.I. Cogolludo Agustín and M. González Villa for their help while preparing this manuscript.

CHAPTER 1

Motivic integration

In this chapter we recall several results from [**15, 16, 17, 18**]. We refer to these papers for the proofs of such results. In the first two sections we work over a field k of characteristic zero.

1. Grothendieck ring of varieties

The Grothendieck ring of algebraic varieties over k, denoted by $K_0(\mathrm{Var}_k)$, is the free Abelian group on isomorphism classes $[X]$ of algebraic varieties X over k subject to the relations $[X] = [X - Y] + [Y]$ where $Y \subset X$ is a closed subvariety of X. The Cartesian product of varieties gives the ring structure. The following properties that we will freely use throughout the paper, hold in $K_0(\mathrm{Var}_k)$.

(1) If $f : Y \to Z$ is a fiber bundle with fiber F which is locally trivial in the Zariski topology, then $[Y] = [F][Z]$.
(2) If a variety X is partitioned by locally closed subvarieties $X_1, \ldots, X_n$, then $[X] = [X_1] + \ldots + [X_n]$.
(3) If $f : Y \to Z$ is a bijective morphism, then $[Y] = [Z]$. The proof of this property is deduced from the proof of the same property at the level of virtual Hodge polynomials which can be found for instance in [**8**].

Let $\mathbb{L} = [\mathbb{A}^1_k]$ denote the class in $K_0(\mathrm{Var}_k)$ of the affine line. The naive motivic ring $\mathcal{M}_k$ of algebraic varieties over k is the polynomial ring $\mathcal{M}_k = K_0(\mathrm{Var}_k)[\mathbb{L}^{-1}]$. Let $F^m\mathcal{M}_k$ denote the subgroup of $\mathcal{M}_k$ generated by $[X]\mathbb{L}^{-i}$ with $\dim X - i \leq -m$ and $\widehat{\mathcal{M}}_k$ denote the completion of $\mathcal{M}_k$ with respect to the filtration $F^{\cdot}$. This completion was first introduced by M. Kontsevich.

2. The arc space of a variety

Let X be a nonsingular irreducible algebraic variety over k of pure dimension d. For any $n \in \mathbb{N}$, let $\mathcal{L}_n(X)$ denote the space of arcs modulo t^{n+1} on X which has a structure of k-variety, whose K-rational points, for any field K containing k, are the $K[t]/(t^{n+1})K[t]$-rational points of X. The arc space $\mathcal{L}(X)$ of X is the projective limit of the algebraic varieties $\mathcal{L}_n(X)$. For any $n \in \mathbb{N}$, let $\pi_n : \mathcal{L}(X) \to \mathcal{L}_n(X)$ be the natural projection. For any arc $\varphi \in \mathcal{L}(X)$, the origin of the arc is $\pi_0(\varphi)$. For any closed point $x \in X$, let $\mathcal{L}_x(X)$ (resp. $\mathcal{L}_{n,x}(X)$) be the arcs (resp. truncated arcs) with origin at x. The above definitions extend to the case where X is a reduced and separated scheme of finite type over $k[t]$. For any $n \in \mathbb{N}$, $\mathcal{L}_n(X)$ is the k-scheme which represents the functor

$$R \mapsto \mathrm{Mor}_{k[t]-schemes}(\mathrm{Spec}\, R[t]/t^{n+1}R[t], X)$$

defined in the category of k-algebras and again $\mathcal{L}(X)$ is its projective limit. The truncation map will be also denoted by $\pi_n : \mathcal{L}(X) \to \mathcal{L}_n(X)$.

Let A be a *semialgebraic*, resp. $k[t]$-*semialgebraic*, subset of $\mathcal{L}(X)$; it is called *stable at level* $n \in \mathbb{N}$ if $A = \pi_n^{-1}\pi_n(A)$. We remark that if A is stable at level n then it is stable at level $n' \geq n$. The set A is called *stable* if it is stable at some level n. A subset $A \in \mathcal{L}(X)$ is *cylindrical at level* n if $A = \pi_n^{-1}(C)$ with C a constructible set, and A is *cylindrical* if it is cylindrical at some level. Denote by $\mathbf{B}^t$ the set of all $k[t]$-semialgebraic subsets of $\mathcal{L}(X)$.

The *motivic measure* on $\mathcal{L}(X)$ is the unique map $\mu_X : \mathbf{B}^t \to \widehat{\mathcal{M}}_k$ such that:

(a) If $A \in \mathbf{B}^t$ is stable at level n, then $\mu_X(A) = [\pi_n(A)]\mathbb{L}^{-(n+1)d}$.
(b) If $A \in \mathbf{B}^t$ is contained in $\mathcal{L}(S)$ with S a reduced closed subscheme of $X \otimes_k k[t]$ with $\dim_{k[t]} S < \dim X$, then $\mu_X(A) = 0$.
(c) Let $A_i \in \mathbf{B}^t$ for all $i \in \mathbb{N}$. Assume that the A_i's are mutually disjoint and that $A := \bigcup_{i\in\mathbb{N}} A_i$ is $k[t]$-semialgebraic. Then $\sum_{i\in\mathbb{N}} \mu_X(A_i)$ converges in $\widehat{\mathcal{M}}_k$ to $\mu_X(A)$.

Because there might exist cylindrical subsets of $\mathcal{L}(X)$ which are not semialgebraic the motivic measure has been extended to a measure, also denoted μ_X, defined over the Boolean algebra of the *measurable* subsets of $\mathcal{L}(X)$, see [**18**, Appendix]. The above properties hold for measurable subsets of $\mathcal{L}(X)$ too.

For a measurable subset A in $\mathcal{L}(X)$ and a function $\alpha : A \to \mathbb{Z} \cup \{\infty\}$, we say that $\mathbb{L}^{-\alpha}$ is *integrable* on A if the fibres of α are measurable, $\alpha^{-1}(\infty)$ has measure zero and the *motivic integral*

$$\int_A \mathbb{L}^{-\alpha} d\mu_X := \sum_{n\in\mathbb{Z}} \mu_X(A \cap \alpha^{-1}(n))\mathbb{L}^{-n} \in \widehat{\mathcal{M}}_k$$

converges in $\widehat{\mathcal{M}}_k$.

DEFINITION 1.1. Let X and Y be k-varieties. A function $\pi : \mathcal{L}(Y) \to \mathcal{L}(X)$ will be call a $k[t]$-*morphism* if it is induced by a morphism of $k[t]$-schemes $Y \otimes_k k[t] \to X \otimes_k k[t]$.

THEOREM 1.2 (Change variables formula). *Let X and Y be smooth k-varieties of pure dimension d. Let $\pi : \mathcal{L}(Y) \to \mathcal{L}(X)$ be a $k[t]$-morphism. Let A and B be $k[t]$-semialgebraic subsets of $\mathcal{L}(X)$ and $\mathcal{L}(Y)$, respectively. Assume that π induces a bijection between B and A. Then, for any function $\alpha : A \to \mathbb{Z} \cup \{\infty\}$ such that $\mathbb{L}^{-\alpha}$ is integrable on A, we have*

$$\int_A \mathbb{L}^{-\alpha} d\mu_X = \int_B \mathbb{L}^{-\alpha\circ\pi - \operatorname{ord}_t J_\pi(y)} d\mu_Y,$$

where $\operatorname{ord}_t J_\pi(y)$, for any $y \in \mathcal{L}(Y)$, denotes the t-order of the Jacobian of π at y.

3. Local Denef-Loeser motivic zeta function

Let $h \in k[[x_1, \ldots, x_d]]$ be a formal power series in the maximal ideal of the formal power series ring. Let $X := \mathbb{A}_k^d$ be the d-dimensional affine space and $x = \mathbf{0}$ its origin. Set $V_n := \{\varphi \in \mathcal{L}_x(X) : \operatorname{ord}(h \circ \varphi) = n\}$. The *local Denef-Loeser motivic zeta function of* h is the power series

$$Z_{DL}(h, T) := \sum_{n\geq 1} \mu_X(V_n)T^n \in \mathcal{M}_k[[T]]. \tag{1.1}$$

Since V_n is a stable semialgebraic set at level n of $\mathcal{L}_{n,x}(X)$, then we have

$$\mu_X(V_n) = [\pi_n(V_n)]\mathbb{L}^{-(n+1)d} \in \mathcal{M}_k.$$

In fact $Z_{DL}(h,T)$ belongs to the subring $\mathcal{N}$ of the ring $\widehat{\mathcal{M}}_k[[T]]$ which is generated by the image in $\widehat{\mathcal{M}}_k[[T]]$ of $\mathcal{M}_k[T]$ and $(1-\mathbb{L}^{-a}T^b)^{-1}$, $a\in\mathbb{N}, b\in\mathbb{P}$. Denef and Loeser in [**17**] introduced the *naive motivic zeta function of* h as

$$Z_{\text{naive}}(h,T):=\sum_{n\geq 1}[\pi_n(V_n)]\mathbb{L}^{-nd}T^n,$$

then $\mathbb{L}^d Z_{DL}(h,T)=Z_{\text{naive}}(h,T)$.

We will work in a slightly more general set up. In order to be able to use the change variables formula we consider $Z_{DL}(h,\omega,T)$ where ω is a regular differential form on X and the pair (h,ω) will satisfy the following condition.

SUPPORT CONDITION 1.3. *The pair* (h,w) *satisfies the support condition if and only if*

(1) $h(\mathbf{x})=\prod_{j=1}^d x_j^{N_j}f(\mathbf{x})$, $N_j\in\mathbb{N}$, *with* x_j *does not divide* f *for any* $j=1,\ldots,d$,
(2) ω *is a regular differential form of type* $\omega=\prod_{j=1}^d x_j^{\nu_j-1}dx_1\wedge\ldots\wedge dx_d$, $\nu_j\geq 1$,
(3) $N_j=0$ *implies* $\nu_j=1$, *for any* $j=1,\ldots,d$.

Set $V_{n,m}:=\{\varphi\in V_n \mid \operatorname{ord}(\prod_{j=1}^{d} x_j^{\nu_j-1}\circ\varphi)=m\}$. For a given n, there are finitely many m such that $V_{n,m}\neq\emptyset$, because of the support condition, cf. [**47**]. The *local Denef-Loeser motivic zeta function of a pair* (h,ω) is the rational function

$$Z_{DL}(h,w,T):=\sum_{n\in\mathbb{N}}\left(\sum_{m\in\mathbb{N}}\mathbb{L}^{-m}\mu_X(V_{n,m})\right)T^n\in\mathcal{N}. \tag{1.2}$$

The local topological zeta function $Z_{\text{top},0}(h,\omega,s)$ is obtained from $Z_{DL}(h,w,T)$ by the following procedure, see [**15**, § 2.3]. First substitute T by $\mathbb{L}^{-s}$ in $Z_{DL}(h,\omega,T)$, then expand $\mathbb{L}^{-s}$ and $(\mathbb{L}-1)(1-\mathbb{L}^{-\nu+Ns})^{-1}$ into series in $\mathbb{L}-1$. Finally take the usual Euler characteristic χ_{top} in étale $\mathbf{Q}_\ell$-cohomology, this works because $\chi_{\text{top}}(\mathbb{L})=1$. Then $Z_{\text{top},0}(h,\omega,s)$ is the rational function

$$Z_{\text{top},0}(h,\omega,s):=\chi_{\text{top}}(Z_{DL}(h,\omega,\mathbb{L}^{-s})). \tag{1.3}$$

We will use the symbol $\chi_{\text{top}}(\bullet(\mathbb{L}^{-s}))$ to denote the composition of the above three operations whenever it makes sense.

REMARK 1.4. See [**18**, Remark 1.19] to generalize the results presented here to schemes over $k[[t]]$ instead over $k[t]$.

CHAPTER 2

Generating functions and Newton polyhedra

1. Generating functions for integer points in rational polyhedra

In this section some well known facts about generating functions of rational polyhedra are reviewed. We use as a reference [**42**, Section 4.6] and [**5**].

Let $\mathbb{R}^d$ be the Euclidean d-space with the standard scalar product $\mathbf{x} \cdot \mathbf{y} = \sum_{l=1}^{d} x_l y_l$. A *rational polyhedron* $P \subset \mathbb{R}^d$ is the set of solutions of a finite system of linear inequalities with integer coefficients:

$$P := \{\mathbf{x} \in \mathbb{R}^d : \boldsymbol{\beta}_i \cdot \mathbf{x} \leq c_i \text{ for } i = 1, \ldots, m\}, \text{ where } \boldsymbol{\beta}_i \in \mathbb{Z}^d \text{ and } c_i \in \mathbb{Z}.$$

A bounded rational polyhedron is called a *polytope*. A nonempty polyhedron is called a *cone* if $\lambda \mathbf{x} \in P$ whenever $x \in P$ and $\lambda \geq 0$. A pointed polyhedral cone is a cone which does not contain a line.

The *algebra of polyhedra* $\mathcal{P}(\mathbb{R}^d)$ is the $\mathbb{Q}$-vector space spanned by the indicator functions $[P]$ of all polyhedra $P \subset \mathbb{R}^d$, where the *indicator function* $[P] : \mathbb{R}^d \to \mathbb{R}$ of P is defined by

$$[P](x) = \begin{cases} 1 & \text{if } x \in P, \\ 0 & \text{if } x \notin P. \end{cases}$$

We will use the same notation $[\bullet]$ for indicator functions and for elements in the Grothendieck ring of algebraic varieties, nevertheless we hope no confusion will arise.

Let $P \subset \mathbb{Q}^d$ be a rational polyhedron, with the set of integral points in P we associate the generating function

$$\Phi_P(\mathbf{x}) := \sum_{\boldsymbol{\alpha} \in P \cap \mathbb{Z}^d} x_1^{\alpha_1} \ldots x_d^{\alpha_d}.$$

These series define a map $\Phi : \mathcal{P}(\mathbb{Q}^d) \to \mathbb{Q}(\mathbf{x})$ with the following properties:

(1) if $P_1, \ldots, P_r \subset \mathbb{R}^d$ are rational polyhedra whose indicator functions satisfy a linear identity $\alpha_1[P_1] + \ldots + \alpha_r[P_r] = 0$, with $\alpha_i \in \mathbb{Q}$, then

$$\alpha_1 \Phi_{P_1}(\mathbf{x}) + \ldots + \alpha_r \Phi_{P_r}(\mathbf{x}) = 0.$$

(2) If $\mathbf{g} + P$ is a translation of P by an integer vector $\mathbf{g} \in \mathbb{Z}^d$ then $\Phi_{\mathbf{g}+P}(\mathbf{x}) = \mathbf{x}^{\mathbf{g}}\, \Phi_P(\mathbf{x})$.

(3) $\Phi_{\{\mathbf{0}\}}(\mathbf{x}) = 1$.

Let $\mathcal{C}$ be a pointed polyhedral cone. The one dimensional faces of $\mathcal{C}$ are called *extreme rays*. A pointed polyhedral cone has only finitely many extreme rays and it is the convex hull of its extreme rays. A *simplicial* cone σ is an e-dimensional pointed convex polyhedral cone with e extreme rays; it may be also defined as a cone generated by e linearly independent integer vectors $\boldsymbol{\beta}_1, \ldots, \boldsymbol{\beta}_e$, thus $\sigma = \{\lambda_1 \boldsymbol{\beta}_1 + \ldots + \lambda_e \boldsymbol{\beta}_e, \lambda_i \in \mathbb{R}_+\}$.

A triangulation of $\mathcal{C}$ consists of a finite collection $\Gamma = \{\sigma_1, ..., \sigma_t\}$ of simplicial cones such that

i) $\bigcup \sigma_i = \mathcal{C}$;
ii) if $\sigma \in \Gamma$, then every face of σ is in Γ;
iii) $\sigma_i \cap \sigma_j$ is a common face of σ_i and σ_j.

It is proved that a pointed convex polyhedral cone $\mathcal{C}$ possesses a triangulation Γ whose 1-dim elements are the extreme rays of $\mathcal{C}$. We will always consider these triangulations in the following.

We are mainly interested in *positive* points in a pointed convex polyhedral cone $\mathcal{C}$. Define $E := \mathcal{C} \cap \mathbb{N}^d$, resp. $\overline{E} := \mathcal{C} \cap \mathbb{P}^d$. Their generating functions are computed using triangulations. The boundary of $\mathcal{C}$, denoted $\partial\mathcal{C}$, is the union of all the facets of $\mathcal{C}$. If Γ is a triangulation of $\mathcal{C}$, let $\partial\Gamma$ denote the set $\{\sigma \in \Gamma, \sigma \in \partial\mathcal{C}\}$, and $\overline{\Gamma} = \Gamma \setminus \partial\Gamma$. Let $\sigma \in \Gamma$ be a simplicial cone, we set $E_\sigma := \sigma \cap \mathbb{N}^d$ and $\overline{E}_\sigma := \{v \in E_\sigma : v \notin E_\tau, \forall \tau \subset \sigma\}$. Then $\overline{E}$ is the disjoint union $\bigcup_{\sigma \in \overline{\Gamma}} \overline{E}_\sigma$ and

$$\Phi_{\overline{E}}(\mathbf{x}) = \sum_{\sigma \in \overline{\Gamma}} \Phi_{\overline{E}_\sigma}(\mathbf{x}).$$

Therefore to compute generating functions of cones, we compute generating functions of simplicial cones. Let $\mathbf{a}_1, ..., \mathbf{a}_t$ be a set of linearly independent integer vectors which generate the cone

$$F := \{\boldsymbol{\lambda} \in \mathbb{N}^d : n\boldsymbol{\lambda} = \lambda_1 \mathbf{a}_1 + \ldots + \lambda_t \mathbf{a}_t,\ n \in \mathbb{N},\ \lambda_i \in \mathbb{N}\},$$

$\mathbf{a}_1, ..., \mathbf{a}_t$ will be called a *set of quasi generators* of F. Define the interior $\overline{F}$ of F as

$$\overline{F} := \{\boldsymbol{\lambda} \in \mathbb{P}^d : n\boldsymbol{\lambda} = \lambda_1 \mathbf{a}_1 + \ldots + \lambda_t \mathbf{a}_t,\ n \in \mathbb{P},\ \lambda_i \in \mathbb{P}\}.$$

Consider the finite set $\overline{D}_F := \{\boldsymbol{\lambda} \in F : \boldsymbol{\lambda} = \lambda_1 \mathbf{a}_1 + \ldots + \lambda_t \mathbf{a}_t,\ 0 < \lambda_i \leq 1\}$. For any $\boldsymbol{\lambda} \in \overline{F}$, there exist unique $\boldsymbol{\beta} \in \overline{D}_F$ and $\lambda_1, \ldots, \lambda_t \in \mathbb{N}$ such that $\boldsymbol{\lambda} = \boldsymbol{\beta} + \lambda_1 \mathbf{a}_1 + \ldots + \lambda_t \mathbf{a}_t$. Then [**42**, Prop. 4.6.8] yields:

$$\Phi_{\overline{F}}(\mathbf{x}) = \frac{\left(\sum_{\boldsymbol{\beta} \in \overline{D}_F} \mathbf{x}^{\boldsymbol{\beta}}\right)}{\prod_{i=1}^{t} (1 - \mathbf{x}^{\mathbf{a}_i})}. \tag{2.1}$$

Given F, there is a unique set $CF(\overline{F}) := \{\boldsymbol{\beta}_1, ..., \boldsymbol{\beta}_t\}$ of primitive quasi generators. We say that $\overline{F}$ is strictly generated by $\boldsymbol{\beta}_1, ..., \boldsymbol{\beta}_t$ and call

$$G_{\overline{F}} := \{\boldsymbol{\lambda} \in \mathbb{P}^d : \boldsymbol{\lambda} = \lambda_1 \boldsymbol{\beta}_1 + \ldots + \lambda_t \boldsymbol{\beta}_t,\ 0 < \lambda_i \leq 1\}$$

the fundamental set of $\overline{F}$.

If $\mathcal{C}$ is a pointed polyhedral cone and Γ a triangulation of $\mathcal{C}$, from [**42**, Prop. 4.6.10], we know that $CF(\overline{E}) := \bigcup_{\sigma \in \overline{\Gamma}} CF(\overline{F}_\sigma)$ is the set of $\boldsymbol{\beta} \in \mathcal{C} \cap \mathbb{P}^d$, which lie on extreme rays of $\mathcal{C}$ such that $\boldsymbol{\beta} \neq n\boldsymbol{\beta}'$ for some $n > 1$ and $\boldsymbol{\beta}' \in \mathcal{C} \cap \mathbb{P}^d$. Furthermore in [**42**, Theorem 4.6.11] it is proved that the rational function $\Phi_{\overline{E}}(\mathbf{x}) \in \mathbb{Q}[\mathbf{x}][D(\mathbf{x})^{-1}]$ where

$$D(\mathbf{x}) = \prod_{\boldsymbol{\beta} \in CF(\overline{E})} (1 - \mathbf{x}^{\boldsymbol{\beta}}). \tag{2.2}$$

The last result we need is the following. Let $\mathcal{C}$ be a pointed polyhedral cone, let $b_1, ..., b_d \in \mathbb{Z}$ such that for each $r \in \mathbb{N}$, the number $g(r)$ of points in $\overline{E} = \mathcal{C} \cap \mathbb{P}^d$ such that $b_1\mathbf{a}_1 + ... + \mathbf{a}_d b_d = r$ is finite. Let $G(\lambda) = \sum_{r\in\mathbb{N}} g(r)\lambda^r$. Then

$$G(\lambda) = \Phi_{\overline{E}}(\lambda^{b_1}, ..., \lambda^{b_d}). \tag{2.3}$$

2. Motivic zeta function and Newton polyhedra

NEWTON DIAGRAMS FOR SERIES. Let $h = \sum_{\mathbf{n}\in\mathbb{N}^d} a_{\mathbf{n}}\mathbf{x}^{\mathbf{n}} \in k[[\mathbf{x}]]$ be a formal power series with $h(\mathbf{0}) = 0$. The *support* of h is the set $\operatorname{supp}(h) = \{\mathbf{n} \in \mathbb{N}^d : a_{\mathbf{n}} \neq 0\}$. The *Newton polyhedron* $\Gamma(h)$ of h is the convex hull in $\mathbb{R}^d_+$ of the set $\bigcup_{\mathbf{n}\in\operatorname{supp}(h)}(\mathbf{n} + (\mathbb{R}_+)^d)$. The *Newton polytope* or *Newton diagram* $ND(h)$ of h is the union of all compact faces of $\Gamma(h)$; the set of all compact faces is denoted by $CF(h)$. The *principal part* of h is the polynomial $h|_{ND(h)} := \sum_{\mathbf{n}\in ND(h)} a_{\mathbf{n}}x^{\mathbf{n}}$. For any $\tau \in CF(h)$ we denote by h_τ the polynomial $\sum_{\mathbf{n}\in\tau} a_{\mathbf{n}}x^{\mathbf{n}}$. The principal part of h is called *non-degenerate* if for each closed proper face $\tau \in CF(h)$, the subscheme of $\mathbf{G}^d_{m,k}$ defined by

$$\frac{\partial h_\tau}{\partial x_1} = \dots = \frac{\partial h_\tau}{\partial x_d} = 0$$

is empty.

NEWTON DIAGRAMS FOR POLYNOMIALS. Let $h : \mathbb{A}^d_k \to \mathbb{A}^1_k$ be a regular morphism, $h(\mathbf{x}) = \sum_{\mathbf{n}\in\mathbb{N}^d} a_{\mathbf{n}}\mathbf{x}^{\mathbf{n}}$. The *support* of h is the set $\operatorname{supp}(h) = \{\mathbf{n} \in \mathbb{N}^d : a_{\mathbf{n}} \neq 0\}$. The *global Newton polytope* $\Gamma_{\mathrm{gl}}(h)$ of h is the convex hull in $\mathbb{R}^d_+$ of the set $\operatorname{supp}(h)$. The polynomial h is called 0-*non-degenerate* if for each closed face $\tau \subset \Gamma_{\mathrm{gl}}(h)$, including $\tau = \Gamma_{\mathrm{gl}}(h)$, the subscheme of $\mathbf{G}^d_{m,k}$ defined by $h_\tau = 0$ is smooth over k.

The *Newton polyhedron* $\Gamma_\infty(h)$ *of h at infinity* is the convex hull of $\operatorname{supp}(h) \cup \{0\}$. The polynomial h is *non-degenerate with respect to* $\Gamma_\infty(h)$ if for every face τ of $\Gamma_\infty(h)$ (of any dimension), which does not contain the origin, the subscheme of $\mathbf{G}^d_{m,k}$ defined by

$$\frac{\partial h_\tau}{\partial x_1} = \dots = \frac{\partial h_\tau}{\partial x_d} = 0$$

is empty. If k is algebraically closed then the subscheme of $\mathbf{G}^d_{m,k}$ defined by $h_\tau = 0$ is smooth over k if and only if $h_\tau, \frac{\partial h_\tau}{\partial x_1}, \dots, \frac{\partial h_\tau}{\partial x_d}$ have no common zero on the torus $\mathbf{G}^d_{m,k}$.

Let $h \in k[[\mathbf{x}]]$ be a formal power series and let ω be a regular differential form such that (h,ω) satisfies the support condition 1.3. Assume that $h(\mathbf{x}) = \prod_{j=1}^d x_j^{N_j} f(\mathbf{x})$, $N_j \in \mathbb{N}$, where x_j does not divide f for any $j = 1, \dots, d$, and the form ω equals $\prod_{j=1}^{d} x_j^{\nu_j - 1} dx_1 \wedge \dots \wedge dx_d$, $\nu_j \geq 1$.

We recall more known definitions and properties. For $\mathbf{k} = (k_1, \dots, k_d) \in \mathbb{R}^d_+$, we define $m_h(\mathbf{k}) := \inf_{\mathbf{x}\in\Gamma(h)}\{\mathbf{k}\cdot\mathbf{x}\}$ and $\sigma_\omega(\mathbf{k}) := \nu_1 k_1 + \dots + \nu_d k_d$. Since h is obtained from f multiplying by a monomial then $\Gamma(h)$ is a translation of $\Gamma(f)$. In particular $m_h(\mathbf{k}) = m_f(\mathbf{k}) + N_1 k_1 + \dots + N_d k_d$.

The first meet locus of $\mathbf{k} \in \mathbb{R}^d_+$ is $F(\mathbf{k}) := \{\mathbf{x} \in \Gamma(h) : \mathbf{k}\cdot\mathbf{x} = m_h(\mathbf{k})\}$. If τ is a face of $\Gamma(h)$ (or $\Gamma(f)$) the cone associated with τ is the convex polyhedral cone, in the dual space, defined by $\Delta_\tau := \{\mathbf{k} \in \mathbb{R}^d_+ : F(\mathbf{k}) = \tau\}$.

It is well-known that the cones associated with the elements of $CF(h)$ give a partition of (the dual space) $\mathbb{R}^d_{>0}$ in a disjoint union $\bigcup_{\tau\in CF(h)}\Delta_\tau$. It turns out that for each $\mathbf{k}=(k_1,\ldots,k_d)\in\mathbb{P}^d$ there exists a unique compact face τ such that $\mathbf{k}\in\Delta_\tau$. Given $\tau\in CF(h)$ we define, see [**12**] for a similar definition in the p-adic case,

$$S_{\Delta_\tau}(h,\omega,T) := \sum_{\mathbf{k}\in\mathbb{P}^d\cap\Delta_\tau} \mathbb{L}^{-\sigma_\omega(\mathbf{k})}T^{m_h(\mathbf{k})}. \tag{2.4}$$

Using the recalled results on generating functions, in particular (2.3), one has

$$S_{\Delta_\tau}(h,\omega,T) = \Phi_{\mathbb{P}^d\cap\Delta_\tau}(\mathbb{L}^{-\nu_1}T^{p_1},\ldots,\mathbb{L}^{-\nu_d}T^{p_d}), \tag{2.5}$$

where $(p_1,\ldots,p_d)\in\tau$, for instance one of its vertices. In what follows we write $\Phi_{\Delta_\tau}(\mathbf{x}) := \Phi_{\mathbb{P}^d\cap\Delta_\tau}(\mathbf{x})$. The term $S_{\Delta_\tau}(h,\omega,T)$ can be computed as follows. Take a partition of the cone Δ_τ into rational simplicial cones Δ_i, $i=1,\ldots,s$, then $S_{\Delta_\tau}(h,\omega,T)=\sum_{i=1}^s S_{\Delta_i}(h,\omega,T)$. If Δ_i is the cone strictly generated by linearly independent vectors $\mathbf{a}_1,\ldots,\mathbf{a}_r\in\mathbb{N}^d$ then (2.1) implies

$$S_{\Delta_i}(h,\omega,T) = \left(\sum_{\mathbf{g}\in G_i}\mathbb{L}^{-\sigma_\omega(\mathbf{g})}T^{m_h(\mathbf{g})}\right)\prod_{j=1}^r \frac{1}{1-\mathbb{L}^{-\sigma_\omega(\mathbf{a}_j)}T^{m_h(\mathbf{a}_j)}}, \tag{2.6}$$

where G_i is the fundamental set of $\Delta_i\cap\mathbb{P}^d$:

$$G_i := \mathbb{N}^d\cap\left\{\sum_{j=1}^r \mu_j\mathbf{a}_j \;\middle|\; 0<\mu_j\le 1 \text{ for } j=1,\ldots,r\right\}.$$

The multiplicity $\operatorname{mult}(\Delta_i)$ of Δ_i is the cardinality of G_i. It is also equal to the volume of the parallelepiped spanned by $\mathbf{a}_1,\ldots,\mathbf{a}_r$ with respect to the volume form $\tilde{\omega}$ on the vector space V generated by $\{\mathbf{a}_1,\ldots,\mathbf{a}_r\}$ normalized such that the parallelepiped spanned by a lattice basis of $\mathbb{Z}^d\cap V$ has volume 1. We define

$$J_{\Delta_i}(h,\omega,s) := \frac{\operatorname{mult}(\Delta_i)}{\prod(\sigma_\omega(\mathbf{a}_j)+m_h(\mathbf{a}_j)s)}. \tag{2.7}$$

If Δ_τ is a r-dimensional rational convex cone and $\Delta_\tau=\bigcup\Delta_i$ is a decomposition in rational simplicial cones Δ_i of dimension r such that $\dim(\Delta_i\cap\Delta_j)<r$ for $i\neq j$, then we define $J_{\Delta_\tau}(h,\omega,s) := \sum J_{\Delta_i}(h,\omega,s)$. In fact $J_{\Delta_\tau}(h,\omega,s)$ is $J_{\Delta_\tau}(h,\omega,s) = \chi_{\mathrm{top}}((\mathbb{L}-1)^r S_{\Delta_\tau}(h,\omega,\mathbb{L}^{-s}))$ and it does not depend on the decomposition.

The following lemma follows from (2.2), (2.6) and the main theorem in [**11**].

LEMMA 2.1. *$S_{\Delta_\tau}(h,\omega,T)\in\mathbb{Z}[\mathbb{L},\mathbb{L}^{-1},(1-\mathbb{L}^{-\sigma_\omega(\mathbf{a})}T^{m_h(\mathbf{a})})^{-1}][T]$, where $\mathbf{a}$ belongs to the set of vectors such that $\mathbf{a}\cdot\mathbf{x}=M$ is a reduced integral equation of an affine hyperplane containing $\tau\in CF(h)$.*

We denote by N_τ the subvariety of $\mathbf{G}^d_{m,k}$ defined by $\{h_\tau=0\}$, that is

$$N_\tau := \{h_\tau=0\}\cap\mathbf{G}^d_{m,k}$$

and the symbol $[N_\tau]$ means its class in the ring $K_0(\mathrm{Var}_k)$. We also define $L^A_\tau(h) := \mathbb{L}^{-d}\left((\mathbb{L}-1)^d-[N_\tau]\right)\in\mathcal{M}_k$. Since $\Gamma(h)$ is a translation of $\Gamma(f)$, there exists a natural bijection between $CF(h)$ and $CF(f)$. Thus we can also write $[N_\tau] = [\mathbf{G}^d_{m,k}\cap\{f_\tau=0\}]$. Let $\tau\in CF(h)$ with $\dim(\tau)=d-r'$, $1\le r'\le d$.

If $r' = d$, it means τ is a vertex of $\Gamma(h)$, then $[N_\tau] = 0$, Δ_τ is a d-dimensional rational convex polyhedron and

$$L_\tau^A(h) = \mathbb{L}^{-d}(\mathbb{L}-1)^d.$$

In such a case
(2.8)

$$\chi_{\text{top}}\left(L_\tau^A(h)S_{\Delta_\tau}(h,\omega,\mathbb{L}^{-s})\right) = \chi_{\text{top}}\left((\mathbb{L}-1)^d S_{\Delta_\tau}(h,\omega,\mathbb{L}^{-s})\right) = J_{\Delta_\tau}(h,\omega,s).$$

If $r' \in \{1,\dots,d-1\}$, then $h_\tau(\mathbf{x})$ is a weighted homogeneous polynomial with more than one monomial. The quasi-projective variety $W := \mathbb{A}_k^d \setminus \{h_\tau(\mathbf{x}) = 0\}$ is k-isomorphic to the affine algebraic variety $Y := \{(\mathbf{x},z) \in \mathbb{A}_k^d \times \mathbb{A}_k^1 \,:\, zh_\tau(\mathbf{x}) = 1\}$. Under such an isomorphism, $W \cap \mathbf{G}_{m,k}^d$ is isomorphic to $Y \cap \mathbf{G}_{m,k}^{d+1}$. It implies that $[N_\tau] = [\mathbf{G}_{m,k}^d] - [W \cap \mathbf{G}_{m,k}^d] = (\mathbb{L}-1)^d - [Y \cap \mathbf{G}_{m,k}^{d+1}]$ in $K_0(\mathrm{Var}_k)$. Thus

$$L_\tau^A(h) = \mathbb{L}^{-d}[Y \cap \mathbf{G}_{m,k}^{d+1}].$$

LEMMA 2.2 (see e.g. [**12**]). *Let $g \in k[x_1,\dots,x_d]$. If $d - r' := \dim \Gamma_\infty(g) < d$ then there exists a coordinate change U on the torus $\mathbf{G}_{m,k}^d$ such that $g(\mathbf{x}) = (g \circ U^{-1})(\mathbf{y}) = \tilde{g}(y_1,\dots,y_{d-r'})$, where $\tilde{g} \in k[y_1,\dots,y_{d-r'},(y_1\cdots y_{d-r'})^{-1}]$.*

Moreover if τ is a face of $\Gamma_\infty(g)$ then $U^{-1}(\tau)$ is a face of $\tilde{g}$ and g is non-degenerate with respect to $\Gamma_\infty(g)$ if and only if $\tilde{g}$ is non-degenerate with respect to $\Gamma_\infty(\tilde{g})$

In our case $g_\tau := zh_\tau(\mathbf{x})$ is a weighted homogeneous polynomial whose Newton polyhedron has dimension $d - r' + 1 < d + 1$. Applying Lemma 2.2 we find a homogeneous polynomial $\tilde{g}_\tau(\mathbf{y})$ in $d-r'+1$ variables such that the variety $Y \cap \mathbf{G}_{m,k}^{d+1}$ is k-isomorphic to $\mathbf{G}_{m,k}^{r'} \times G_\tau^*$ where $G_\tau^* = \{\mathbf{y} \in \mathbf{G}_{m,k}^{d-r'+1} : \tilde{g}_\tau(\mathbf{y}) = 1\}$. It turns out that $\mathbb{L}^d L_\tau^A(h) = [Y \cap \mathbf{G}_{m,k}^{d+1}] = (\mathbb{L}-1)^{r'}[G_\tau^*]$ in $K_0(\mathrm{Var}_k)$. The convex rational cone Δ_τ has dimension r', thus

$$\chi_{\text{top}}\left(L_\tau^A(h)S_{\Delta_\tau}(h,\omega,\mathbb{L}^{-s})\right) = \chi_{\text{top}}([G_\tau^*])J_{\Delta_\tau}(h,\omega,s). \tag{2.9}$$

Later on we will consider germs which are degenerate with respect to their Newton polyhedra. In the following definition we collect the terms corresponding to the A-part in the decomposition of the arc space according to these polyhedra.

DEFINITION 2.3. Let $h \in k[[\mathbf{x}]]$ be a power series and ω a differential form such that (h,ω) satisfies condition (1.3). The A-part or the part corresponding to the Newton polyhedron of h is defined by the rational function

$$Z_{DL}^A(h,\omega,T) := \sum_{\tau \in CF(h)} L_\tau^A(h)\, S_{\Delta_\tau}(h,\omega) \in \mathcal{M}_k[(1-\mathbb{L}^{-\sigma_\omega(\mathbf{a})}T^{m_h(\mathbf{a})})^{-1}][T],$$

where $\mathbf{a}$ belongs to the set of vectors such that $\mathbf{a}\cdot\mathbf{x} = M$ is a reduced integer equation of an affine hyperplane containing some $\tau \in CF(h)$. In the same way we define the A-part of (h,ω) for the local topological zeta function as the rational function

$$Z_{\text{top},0}^A(h,\omega,s) = \sum_{\tau \in CF(h)} \chi_{\text{top}}([G_\tau^*])J_{\Delta_\tau}(h,\omega,s),$$

where $\chi_{\text{top}}([G_\tau^*]) := 1$ if $\dim(\tau) = 0$.

If h has non-degenerate Newton principal part, $Z_{DL}(h,\omega,T)$ is written in terms of some invariants of $CF(h)$ and ω. This formula has been proved by J. Denef and K. Hoornaert in the p-adic setting, [**12**]. Our proof is based on their results, so we follow their notation too. In fact we provide the proof because it will give some light in the arc decomposition that we have to do in the quasi-ordinary case. G. Guibert has communicated to the authors that he has recently obtained a similar formula, see [**24**].

THEOREM 2.4. *Let* $h(\mathbf{x}) = \prod_{j=1}^{d} x_j^{N_j} f(\mathbf{x})$, $N_j \in \mathbb{N}$, *be a regular function with* h *non-degenerate with respect to (all the compact faces of) its Newton polyhedron and let* $\omega = \prod_{j=1}^{d} x_j^{\nu_j - 1} dx_1 \wedge \ldots \wedge dx_d$, $\nu_j \geq 1$ *be a differential form satisfying the support condition 1.3. Then*

$$Z_{DL}(h,\omega,T) = \sum_{\tau \in CF(h)} L_\tau(h) S_{\Delta_\tau}(h,\omega,T)$$

where $L_\tau(h) := L_\tau^A(h) + L_\tau^B(h)$ *and* $L_\tau^B(h) := \mathbb{L}^{-d}(\mathbb{L}-1)[N_\tau]\frac{\mathbb{L}^{-1}T}{1-\mathbb{L}^{-1}T}$.

We break the proof in several steps.

Step 1. Classifying arcs.

Let S be the affine hypersurface $x_1 \cdots x_d = 0$. Since S has dimension less than d then by property (b) of the motivic measure the set $\mathcal{L}_0(S)$ has measure zero. Then we only consider arcs $\varphi \in \mathcal{L}_0(\mathbb{A}_k^d) \setminus \mathcal{L}_0(S)$, i.e., $\varphi = (\varphi_1, \ldots, \varphi_d)$ where $\varphi_i(t) = a_{k_i} t^{k_i}$ + higher degree terms. As usual we write $\mathbf{k}(\varphi) := (k_1, \ldots, k_d) \in \mathbb{P}^d$ and $\mathbf{a}(\varphi) := (a_{k_1}, \ldots, a_{k_d}) \in \mathbf{G}_{m,k}^d$.

Let τ be the unique compact face of $ND(h)$ such that $\mathbf{k} = \mathbf{k}(\varphi) \in \Delta_\tau \cap \mathbb{P}^d$. If $\mathbf{a} = \mathbf{a}(\varphi) \in \mathbf{G}_{m,k}^d \setminus N_\tau$ then $\mathrm{ord}(h \circ \varphi) = \mathrm{ord}(h_\tau \circ \varphi) = m_h(\mathbf{k})$. Otherwise, if $\mathbf{a} \in N_\tau$ then $\mathrm{ord}(h \circ \varphi) > m_h(\mathbf{k})$.

The set $\overline{V}_{n,m} = \{\varphi \in \mathcal{L}_0(\mathbb{A}_k^d) \setminus \mathcal{L}_0(S) : \mathrm{ord}(h \circ \varphi) = n,\ \mathrm{ord}(\prod_{j=1}^{d} x_j^{\nu_j - 1} \circ \varphi) = m\}$ can be decomposed as $\overline{V}_{n,m} = \bigcup_{\tau \in CF(f)} \left(V_{n,A,m}^\tau \cup V_{n,B,m}^\tau\right)$, where

$$V_{n,A,m}^\tau := \{\varphi \in \overline{V}_{n,m} : \mathbf{k}(\varphi) \in \Delta_\tau,\ \mathbf{a}(\varphi) \in \mathbf{G}_{m,k}^d \setminus N_\tau\},$$

$$V_{n,B,m}^\tau := \{\varphi \in \overline{V}_{n,m} : \mathbf{k}(\varphi) \in \Delta_\tau,\ \mathbf{a}(\varphi) \in N_\tau\}.$$

Rewrite the formula (1.2) as:

$$Z_{DL}(h,w,T) = \sum_{\tau \in CF(h)} \left(\sum_{n \geq 1} \left(\sum_{m \in \mathbb{P}} \mathbb{L}^{-m} \mu_X(V_{n,A,m}^\tau) \right) T^n + \right. \tag{2.10}$$
$$\left. + \sum_{n \geq 1} \left(\sum_{m \in \mathbb{P}} \mathbb{L}^{-m} \mu_X(V_{n,B,m}^\tau) \right) T^n \right).$$

Step 2. Computation of the A-part of the series.

Let $\tau \in CF(h)$ and $\mathbf{k} \in \Delta_\tau$. Define

$$V_A^{\mathbf{k}} := \{\varphi \in \mathcal{L}_0(\mathbb{A}_k^d) \setminus \mathcal{L}_0(S) : \mathbf{k}(\varphi) = \mathbf{k},\ f_\tau(\mathbf{a}(\varphi)) \neq 0\}.$$

Given $\varphi \in V_A^{\mathbf{k}}$ then $\operatorname{ord}(h \circ \varphi)$ and $\operatorname{ord}(\prod_{j=1}^{d} x_j^{\nu_j - 1} \circ \varphi)$ depend only on $\mathbf{k}$, therefore $V_A^{\mathbf{k}} \subset V^{\tau}_{n_{\mathbf{k}},A,m_{\mathbf{k}}}$ where $n_{\mathbf{k}} := m_h(\mathbf{k}) = m_f(\mathbf{k}) + \sum_{i=1}^{d} N_i k_i$ and $m_{\mathbf{k}} := \sigma_\omega(\mathbf{k}) - \sum_{i=1}^{d} k_i$. The stable semialgebraic set $V_A^{\mathbf{k}}$ are stable at some level $n'_{\mathbf{k}} \geq n_{\mathbf{k}}$. Thus it has measure $\mu_X(V_A^{\mathbf{k}}) = [\pi_{n'_{\mathbf{k}}}(V_A^{\mathbf{k}})]\mathbb{L}^{-(n'_{\mathbf{k}}+1)d}$. Since

$$[\pi_{n'_{\mathbf{k}}}(V_A^{\mathbf{k}})] = [\mathbf{G}^d_{m,k} \setminus \{f_\tau = 0\}]\mathbb{L}^{dn'_{\mathbf{k}} - k_1 - \ldots - k_d}$$

then $\mathbb{L}^{-m_{\mathbf{k}}}\mu_X(V_A^{\mathbf{k}}) = \mathbb{L}^{-d}\left((\mathbb{L}-1)^d - [N_\tau]\right)\mathbb{L}^{-\sigma_\omega(\mathbf{k})}$.

It turns out that

$$\sum_{n \geq 1}\left(\sum_{m \in \mathbb{P}} \mathbb{L}^{-m}\mu_X(V^{\tau}_{n,A,m})\right)T^n = \sum_{\mathbf{k} \in \mathbb{P}^d \cap \Delta_\tau} \mathbb{L}^{-m_{\mathbf{k}}}\mu_X(V_A^{\mathbf{k}})T^{n_{\mathbf{k}}} =$$

$$= \sum_{\mathbf{k} \in \mathbb{P}^d \cap \Delta_\tau} \mathbb{L}^{-d}\left((\mathbb{L}-1)^d - [N_\tau]\right)\mathbb{L}^{-\sigma_\omega(\mathbf{k})}T^{m(\mathbf{k}) + N_1 k_1 + \ldots + N_d k_d} =$$

$$= \mathbb{L}^{-d}\left((\mathbb{L}-1)^d - [N_\tau]\right)\sum_{\mathbf{k} \in \mathbb{P}^d \cap \Delta_\tau} \mathbb{L}^{-\sigma_\omega(\mathbf{k})}T^{m_h(\mathbf{k})} = L_\tau^A(h) S_{\Delta_\tau}(h, \omega, T).$$

Step 3. Computation of the B-part of the series.

Let $\tau \in CF(h)$. As before for any $\mathbf{k} \in \mathbb{P}^d \cap \Delta_\tau$ we define the sets $V^{\mathbf{k}}_{n,B}$ of arcs φ in $V^{\tau}_{n,B}$ such that $\mathbf{k}(\varphi) = \mathbf{k}$. Note that $V^{\mathbf{k}}_{n,B} \subset V^{\tau}_{n,B,m_{\mathbf{k}}}$.

LEMMA 2.5. *Fix* $\mathbf{k} \in \mathbb{P}^d \cap \Delta_\tau$ *as before. Then*

$$\sum_{n \geq 1} \mathbb{L}^{-m_{\mathbf{k}}}\mu_X(V^{\mathbf{k}}_{n,B})T^n = \mathbb{L}^{-d}(\mathbb{L}-1)[N_\tau]\frac{\mathbb{L}^{-1}T}{1 - \mathbb{L}^{-1}T}\mathbb{L}^{-\sigma_\omega(\mathbf{k})}T^{m_h(\mathbf{k})}.$$

The formula in Theorem 2.4 is deduced from the following equality which is a consequence of the above lemma:

$$\sum_{\mathbf{k} \in \mathbb{P}^d \cap \Delta_\tau}\sum_{n \geq 1} \mathbb{L}^{-m_{\mathbf{k}}}\mu_X(V^{\mathbf{k}}_{n,B})T^n = \mathbb{L}^{-d}(\mathbb{L}-1)[N_\tau]\frac{\mathbb{L}^{-1}T}{1 - \mathbb{L}^{-1}T}S_{\Delta_\tau}(h, \omega, T).$$

PROOF OF LEMMA 2.5. Given $\mathbf{k} \in \mathbb{P}^d \cap \Delta_\tau$, consider the $k[t]$-morphism $\pi_{\mathbf{k}} : Y \to \mathbb{A}^d_{k[t]}$ defined by $\pi_{\mathbf{k}}(y_1, \ldots, y_d) = (t^{k_1}y_1, \ldots, t^{k_d}y_d)$, where $Y := \mathbb{A}^d_k$. Let $\varphi \in V^{\mathbf{k}}_{n,B}$ be an arc, then the equalities $\varphi_i(t) = t^{k_i}\psi_i(t)$, $i = 1, \ldots, d$, define a unique arc $\psi(t) := (\psi_1(t), \ldots, \psi_d(t)) \in \mathcal{L}(Y)$ centered at $\psi(0) = \mathbf{a}(\varphi) \in \mathbf{G}^d_{m,k} \cap \{f_\tau = 0\}$. It satisfies $\operatorname{ord}(f \circ \pi_{\mathbf{k}} \circ \psi) = \operatorname{ord}(f \circ \varphi) = n$ and because of the quasi-homogeneity of f_τ:

$$f \circ \pi_{\mathbf{k}}(y_1, \ldots, y_d) = t^{m(\mathbf{k})}(f_\tau(y_1, \ldots, y_d) + tg(y_1, \ldots, y_d)),$$

where the function g has coefficients in $k[t]$. In particular $n = m_f(\mathbf{k}) + n'$ with $n' := \operatorname{ord}(f_1 \circ \psi) \geq 1$ where $f_1(y_1, \ldots, y_d) := f_\tau(y_1, \ldots, y_d) + tg(y_1, \ldots, y_d)$. The algebraic set $f_\tau(y_1, \ldots, y_d) = 0$ is nonsingular at points of $\mathbf{G}^d_{m,k} \cap \{f_\tau = 0\}$ because of the non-degeneracy condition. In particular at any of such points there are coordinates such that $f_\tau = \bar{y}_1$ and in the same way at such a point $f_1 = 0$ will be non-singular. Applying the change variables formula the proof of the lemma is finished. Remark that the pull-back by $\pi_{\mathbf{k}}$ of the regular differential form $dx_1 \wedge \ldots \wedge dx_d$ is $t^{\sigma_\omega(\mathbf{k})}dy_1 \wedge \ldots \wedge dy_d$. □

A formula for $Z_{\mathrm{top},0}(h,\omega,s)$ for a non-degenerate function h was given in [**14**]. We apply Theorem 2.4 to reprove that formula. We need several standard definitions.

DEFINITION 2.6. Let γ be the convex hull in $\mathbb{R}^d$ of some subset of $\mathbb{Z}^d$. The *volume form* ω_γ *on the affine space* $\mathrm{Aff}(\gamma)$ generated by γ is defined such that the parallelepiped generated by a base of the lattice $\mathrm{Aff}(\gamma)\cap\mathbb{Z}^d$ has volume 1. For any $\tau\in CF(h)$, $V(\tau)$ is the volume of τ relative to the induced lattice; i.e. to the volume form ω_τ.

THEOREM 2.7. [**14**] *If h has non-degenerate principal part and the pair (h,ω) satisfies the support condition (1.3) then*

$$Z_{\mathrm{top},0}(h,\omega,s)=\sum_{\tau\in CF(h),\dim\tau=0} J_{\Delta_\tau}(h,\omega,s)$$
$$+\left(\frac{s}{s+1}\right)\sum_{\tau\in CF(h),\dim\tau\geq 1}(-1)^{\dim(\tau)}\dim(\tau)!V(\tau)J_{\Delta_\tau}(h,\omega,s).$$

The proof follows the same ideas as in equations (2.8) and (2.9). Let $\tau\in CF(h)$. If τ is zero dimensional we have the identity (2.8). Otherwise, following the same notation as in equation (2.9), if $\dim(\tau)=d-r'$ then

$$L_\tau(h)=\mathbb{L}^{-d}\left((\mathbb{L}-1)^{d+1}\frac{\mathbb{L}^{-1}T}{1-\mathbb{L}^{-1}T}+(\mathbb{L}-1)[Y\cap\mathbf{G}^{d+1}_{m,k}]\frac{1-T}{1-\mathbb{L}^{-1}T}\right).$$

Since $d-r'<d$, the denominator of $S_{\Delta_\tau}(h,\omega,T)$ has at most $d-1$ factors and

$$\chi_{\mathrm{top}}\left(\mathbb{L}^{-d}(\mathbb{L}-1)^{d+1}\frac{\mathbb{L}^{-1}T}{1-\mathbb{L}^{-1}T}S_{\Delta_\tau}(h,\omega,\mathbb{L}^{-s})\right)=0.$$

Therefore we have again the term $[Y\cap\mathbf{G}^{d+1}_{m,k}]$. Applying Lemma 2.2 to $g_\tau:=zh_\tau(\mathbf{x})$ we have $[Y\cap\mathbf{G}^{d+1}_{m,k}]=(\mathbb{L}-1)^{r'}[G^*_\tau]$ in $K_0(\mathrm{Var}_k)$ where $G^*_\tau=\{\mathbf{y}\in\mathbf{G}^{d-r'+1}_{m,k}:\tilde{g}_\tau(\mathbf{y})=1\}$ for some non-degenerate weighted homogeneous polynomial $\tilde{g}_\tau(\mathbf{y})$. Finally we also need the following result.

LEMMA 2.8. [**6, 13**] *Let $g(y_1,\ldots,y_n)$ be a 0-non-degenerate polynomial and let $Z^*=\{y\in\mathbf{G}^n_{m,k}:g(\mathbf{y})=0\}$. Then the ℓ-adic Euler characteristic $\chi_{\mathrm{top}}(Z^*)$ is given by*

$$\chi_{\mathrm{top}}(Z^*)=(-1)^{n-1}n!\,\mathrm{Vol}_n(\Delta(g)).$$

Here $\mathrm{Vol}_n(\Delta(g))$ is the n-dimensional Euclidean volume.

In our case the polynomial $g=\tilde{g}_\tau-1$ is 0-non-degenerate and satisfies the hypothesis of the Lemma 2.8, then $\chi_{\mathrm{top}}(G^*)=(-1)^r(r+1)!\,\mathrm{Vol}_{r+1}(\Delta(g))$. Since $\Delta(g)$ is a cone over the origin, $(r+1)!\,\mathrm{Vol}_{r+1}(\Delta(g))=r!V(\tau)$ (this last volume as in definition 2.6). Hence we get the formula in Theorem 2.7.

REMARK 2.9. All the results in this section are also valid in the complex analytic set up.

EXAMPLE 2.10. It is clear that the poles of $Z_{\mathrm{top},0}(h,\omega,s)$ are included in the poles of $Z_{DL}(h,\omega,T)$ but in general the other way around is not true. Consider for instance $h=x_1^3+x_2^3+x_3^3+x_4^3+x_5^6$ and the differential form $\omega=d\mathbf{x}$. This example has been recently studied by Ishii-Kollar in [**27**] to disprove a J. Nash conjecture.

Then h has non-degenerate principal part and its Newton polyhedron has only one compact facet τ which, in the dual, corresponds with the extreme ray $v_\tau :=$ $(2,2,2,2,1)$. Take the vertex $(3,0,0,0,0)$ of the face τ, then

$$S_{\Delta_\tau}(h,\omega,T) = \Phi_{\Delta_\tau}(\mathbb{L}^{-1}T^3, \mathbb{L}^{-1}T^0, \mathbb{L}^{-1}T^0, \mathbb{L}^{-1}T^0, \mathbb{L}^{-1}T^0) = \frac{\mathbb{L}^{-9}T^6}{1-\mathbb{L}^{-9}T^6}.$$

After Lemma 2.1, $(1-\mathbb{L}^{-9}T^6)$ is a possible factor of the denominator of $Z_{DL}(h,\omega,T)$. In fact in this case it is. To prove it, one considers the Igusa p-adic zeta function of h and applies [**12**, Theorem 5.17] which guarantees that $t_0 = -3/2$ is one of the poles for $p \gg 0$. Nevertheless the local topological zeta function is $Z_{\text{top},0}(h,\omega,s) = 1/(1+s)$. This gives an example where poles of the p-adic and the local topological zeta functions are not the same.

CHAPTER 3

Quasi-ordinary power series

In this chapter we recall some known properties of the quasi-ordinary power series. We give the necessary details to describe the method we will use to compute their local zeta functions. Let k be an algebraically closed field of characteristic zero.

Let $h \in k[[\mathbf{x}]][z]$ be a z-polynomial of degree s with coefficients in the formal power series $k[[\mathbf{x}]]$, $\mathbf{x} = (x_1, \dots, x_d)$. Assume that $h(\mathbf{x}, z) = x_1^{N_1} \dots x_d^{N_d} g(\mathbf{x}, z)$ where $N_l \geq 0$ and x_l doesn't divide $g(\mathbf{x}, z)$, $\forall l, l = 1, \dots, d,$.

The power series h is called *quasi-ordinary* (or *k-quasi-ordinary* if we want to emphasize the base field) if its z-discriminant $D_z(g)$ (or equivalently $D_z(h)$) is

$$D_z(g) = x_1^{\alpha_1} x_2^{\alpha_2} \dots x_d^{\alpha_d} \varepsilon(\mathbf{x}),$$

where $\varepsilon(\mathbf{0}) \neq 0$ and $\alpha_i \in \mathbb{N}$. The condition on the z-discriminant implies that $g(\mathbf{x}, z)$ is squarefree in the ring $k[[\mathbf{x}, z]]$. Since q is quasi-ordinary one can write $g(\mathbf{0}, z)$ as $g(\mathbf{0}, z) = u \prod (z - z_b)^{s_b}$, with $s = \sum s_b$, $u \in \mathbf{G}_{m,k}$ and $z_b \in \mathbb{A}_k^1$.

For each root z_b of $g(\mathbf{0}, z)$, the Jung-Abhyankar Theorem, [**1, 36**], states that there are exactly s_b distinct roots of $g(\mathbf{x}, z)$ centred at z_b in $k[[x_1^{1/m_b}, \dots, x_d^{1/m_b}]]$, for some $m_b \in \mathbb{P}$. Thus for each $j \in \{1, \dots, s_b\}$, there exists a fractional power series $\zeta_{b_j} \in k[[x_1^{1/m_b}, \dots, x_d^{1/m_b}]]$ with $\zeta_{b_j}(\mathbf{0}) = z_b$ such that

$$g(\mathbf{x}, z) = \prod_b \prod_{j=1}^{s_b} \left(z - \zeta_{b_j}(x_1^{1/m_b}, \dots, x_d^{1/m_b}) \right). \tag{3.1}$$

In fact, for each root z_b, the product $\prod_{j=1}^{s_b} \left(z - \zeta_{b_j}(x_1^{1/m_b}, \dots, x_d^{1/m_b}) \right)$ is a well defined element in $k[[\mathbf{x}]][z]$. Moreover, this power series is quasi-ordinary too because of the properties of the discriminant.

In principle we are interested just in one of the roots, say $z_b = 0$, otherwise we can make a translation of type $\bar{z} = z - z_b$ to study the corresponding root. Later on when the transversal sections are described we will deal with other roots different from $z_b = 0$, see Chapter 5.

Therefore we may assume

$$h(\mathbf{x}, z) = x_1^{N_1} \dots x_d^{N_d} f(\mathbf{x}, z) u(\mathbf{x}, z), \tag{3.2}$$

where $f(\mathbf{x}, z)$ is a degree n quasi-ordinary z-Weierstrass polynomial, n being the multiplicity of the root $z_b = 0$ in $g(\mathbf{0}, z)$, $u(\mathbf{x}, z)$ is a unity in $k[[\mathbf{x}, z]]$, $u(\mathbf{x}, z)$ being the product of all other roots centred at points different from $z_b = 0$, and $N_l \geq 0$ for any $l = 1, \dots, d$. Therefore, for each $j \in \{1, \dots, n\}$, there exist $\zeta_j \in$

$k[[x_1^{1/m},\dots,x_d^{1/m}]]$ (for some positive integer m) with $\zeta_j(\mathbf{0})=0$ such that

$$f(\mathbf{x},z)=\prod_{j=1}^{n}\left(z-\zeta_j(x_1^{1/m},\dots,x_d^{1/m})\right). \tag{3.3}$$

Let ζ be a root of f, all its conjugates $\tilde{\zeta}_{\boldsymbol{\varepsilon}}:=\zeta(\varepsilon_1 x_1^{1/m},\dots,\varepsilon_d x_d^{1/m})$, where $\varepsilon_l^m=1$, for all $l=1,\dots,d$, are also roots of f. If f is irreducible in $k[[\mathbf{x}]][z]$, then all its roots are conjugate to one of them, say ζ, and the set $\{\tilde{\zeta}_{\boldsymbol{\varepsilon}}\}$ has exactly n distinct elements $\{\zeta=\zeta_1,\dots,\zeta_n\}$ which satisfy

$$f(\mathbf{x},z)=\prod_{p=1}^{n}(z-\zeta_p).$$

The irreducible factors of f in $k[[\mathbf{x}]][z]$ are also k-quasi-ordinary power series. Let $\{f^{(i)}\}_{i\in I}$ be the set of irreducible factors of f. Then

$$f(\mathbf{x},z)=\prod_{i\in I}\prod_{j=1}^{n^{(i)}}(z-\zeta_j^{(i)})$$

The set of indexes I is decomposed in two disjoint subsets: I' and $I\setminus I'$ such that $\#(I\setminus I')\leq 1$, where in $\#(I\setminus I')=1$ if and only if z is an irreducible component of f (i.e. if 0 is root of f).

DEFINITION 3.1. We will say that a variable x_i is *essential* for h if x_i divides $D_z(h)$. The *number of essential variables* will be denoted by $\mathbf{ev}(h)$.

REMARK 3.2. All our results for the local Denef-Loeser motivic zeta function of quasi-ordinary power series are proved in the case where g is reduced. Nevertheless they can be proved in the nonreduced case. This means that g can have multiple components in $k[[\mathbf{x},z]]$. In such a case the power series $h(\mathbf{x},z)$ will be called quasi-ordinary if $x_1^{N_1}\dots x_d^{N_d}g_{\mathrm{red}}(\mathbf{x},z)$ is quasi-ordinary in the above sense.

We are only interested in the local Denef-Loeser motivic zeta function. Since for any arc $\varphi(t)$, with $\varphi(\mathbf{0})=0$, the t-order $\mathrm{ord}(h\circ\varphi)$ does not depend on the unit $u(\mathbf{x},z)$, we usually suppose that $u(\mathbf{x},z)=1$.

1. Characteristic exponents

Given $h(\mathbf{x},z)=\prod_{l=1}^{d}x_l^{N_l}f(\mathbf{x},z)=\prod_{l=1}^{d}x_l^{N_l}\prod_{i\in I}(z-\zeta_i)$, for I indexing the number of roots of f, one has $D_z(f)=\prod(\zeta_k-\zeta_j)$. For two different roots ζ_k and ζ_j of f, using the unique factorization of the discriminant, we have

$$\zeta_k-\zeta_j=x_1^{\lambda_{kj,1}}x_2^{\lambda_{kj,2}}\dots x_d^{\lambda_{kj,d}}\varepsilon_{kj}(x_1^{1/m},\dots,x_d^{1/m}),$$

where $\varepsilon_{kj}(0)\neq 0$, and $\lambda_{kj,t}\in\frac{1}{m}\mathbf{Z}_{\geq 0}$.

DEFINITION 3.3. The set $\Lambda_{CE}(f):=\{\boldsymbol{\lambda}_{kj}\}=\{(\lambda_{kj,1},\dots,\lambda_{kj,d})\}_{kj}\subset\frac{1}{m}\mathbf{Z}_{\geq 0}^d$ will be called *characteristic exponents set* of f. We usually identify $\Lambda_{CE}(h)$ with $\Lambda_{CE}(f)$ because $\Lambda_{CE}(f)$ is the characteristic exponents set of h at the root zero. We usually omit "at the root zero" whenever it is clear.

For each element $\boldsymbol{\lambda}=(\lambda_1,\dots,\lambda_d)\in\Lambda_{CE}(f)$, the corresponding monomial $\mathbf{x}^{\boldsymbol{\lambda}}$ is called *characteristic monomial* of f. In the same way we identify characteristic monomials of h at zero with characteristic monomials of f.

Since the irreducible factors $f^{(i)}$ of f are also quasi-ordinary, the characteristic exponent set $\Lambda_{CE}(f^{(i)})$ has the property that $\Lambda_{CE}(f^{(i)}) \subset \Lambda_{CE}(f)$ for every $i \in I$. The other characteristic exponents of f measure the order of coincidence between distinct irreducible factors of f (as in the case of plane curves).

The following partial ordering will be used in the paper. For any $\boldsymbol{\lambda}, \boldsymbol{\mu} \in \mathbb{Q}^d$, then $\boldsymbol{\lambda} \leq \boldsymbol{\mu}$ if $\lambda_l \leq \mu_l$ for all $l = 1, \ldots, d$. We say that $\boldsymbol{\lambda} < \boldsymbol{\mu}$ if $\boldsymbol{\lambda} \leq \boldsymbol{\mu}$ and there is an l such that $\lambda_l < \mu_l$. If $f(\mathbf{x}, z)$ is irreducible in $k[[\mathbf{x}, z]]$, then the elements of $\Lambda_{CE}(f)$ are totally ordered, see [**32**] and the references therein. But this is no longer true in the non-irreducible case.

Up to now to have the same characteristic exponents set is just that two sets coincide. Later on, we will give a new definition of having the same characteristic exponents (which will extend the usual definition for curves).

2. Newton polyhedron and good coordinates

In this section we show that for every quasi-ordinary power series $h(\mathbf{x}, z)$ there exist coordinates, called *good coordinates*, such that the compact faces of $ND(h)$ have only dimension 1 and 0. This means that the 1-dimensional compact faces of $ND(h)$ are totally ordered by their slopes and they form a *monotone polygonal path*. Our definition is slightly more general than the definition of good coordinates given by P. González in [**20**], [**23**, Lemma 3.16]; we will call his good coordinates *P-good coordinates*. He proves the existence of P-good coordinates for quasi-ordinary power series; in fact his P-good coordinates are a particular case of our good coordinates, see Remark 3.9. The essential difference between both set of coordinates appears in the characteristic exponents which are integers. The advantage of our good coordinated with respect to P-good coordinates is that ours are preserved when passing to transversal sections, see section 5 in this chapter; this is not the case for P-good coordinates.

DEFINITION 3.4. A quasi-ordinary power series $h \in k[[\mathbf{x}]][z]$ is in *good coordinates* if $ND(h)$ is a monotone polygonal path such that either z is a component of h or h_γ is not the product of a monomial in $\mathbf{x}$ and a power of a linear form, where γ is the compact 1-dimensional face of $ND(h)$ which meets the plane $z = 0$, i.e. $h_\gamma \neq x_1^{a_1} \ldots x_d^{a_d}(z - \alpha\, x_1^{b_1} \ldots x_d^{b_d})^a$ for all $\alpha \in \mathbf{G}_{m,k}$.

PROPOSITION 3.5. *There exists a system of coordinates which are good for h.*

PROOF. Since h and f essentially differ by a monomial, the Newton polyhedron $\Gamma(h)$ is obtained from $\Gamma(f)$ by the translation induced by the corresponding monomial. Thus it is enough to prove the result for f being a Weierstrass quasi-ordinary polynomial of degree n. We proceed by induction on n. The case $n = 1$ is clear: we have $f = z + a(\mathbf{x})$, with $a(\mathbf{x}) \in k[[\mathbf{x}]]$, and the change $z_1 = z + a(\mathbf{x})$ is enough.

In order to proceed by induction we need the following notion introduced by Hironaka and studied in [**36**].

DEFINITION 3.6. Let $f(\mathbf{x}, z) = z^n + a_{n-1}(\mathbf{x})z^{n-1} + \ldots + a_0(\mathbf{x})$ be a quasi-ordinary Weierstrass polynomial, set $\tau_0 = (0, \ldots, 0, n)$ the corresponding vertex in $ND(f)$. We say that f is *ν-quasi-ordinary with respect to z* if there is a vertex τ_1 in $ND(f)$, $\tau_1 \neq \tau_0$, such that if τ is the projection of τ_1 over $\mathbb{Q}^d \times \{0\}$ from τ_0 and γ is the segment joining τ_0 and τ then $ND(f)$ is contained in $\bigcup_{\mathbf{n}\in\gamma}(\mathbf{n} + \mathbb{R}_+^{d+1})$ and f_γ is a polynomial which is not a power of a linear form.

EXAMPLE 3.7. The polynomial $f(x,y,z) := z^2(z^2 - xy)^2 + x^3y^5 + x^5y^3$ is ν-quasi-ordinary with respect to z but it is not quasi-ordinary (compute the discriminant).

In [**36**] it is proved that if we make the change $z = z_1 - \frac{a_{n-1}(\mathbf{x})}{n}$ the polynomial f becomes ν-quasi-ordinary. If $ND(f) = \gamma$ (with the above notations) then γ is the only 1-dimensional compact face, therefore the last one, and f_γ is not a power of a linear form so we are done. Otherwise, in [**36**] it is shown that f can be decomposed as a product $f = f_0(\mathbf{x}, z)\, f_1(\mathbf{x}, z)$, $f_i(\mathbf{x}, z) \in k[[\mathbf{x}]][z]$, such that f_γ corresponds with the factor f_0. Recall that the Newton polytope of f is the Minkowski sum of the Newton polytopes of each factor. The factor f_1 is a quasi-ordinary Weierstrass polynomial of degree less than n and we apply induction to conclude. In fact the change of variables we need for the induction step are of type $z_1 = z_2 + b(\mathbf{x})$ and they do not affect the first segment of Newton polytope of f because of the condition $ND(f)$ is contained in $\bigcup_{\mathbf{n}\in\gamma}(\mathbf{n} + \mathbb{R}_+^{d+1})$ in the definition of ν-quasi-ordinary. $\square$

REMARK 3.8. Observe that from the proof of the proposition 3.5 one gets that it is enough to make change of variables of type $z = z_1 + a(\mathbf{x})$, $a(\mathbf{x}) \in k[[\mathbf{x}]]$ to get good coordinates. The procedure to get good coordinates is far from being unique. For instance $f(x,z) = (z^2 - x^5)(z - x^2)(z^3 - x^2)$ is already in good coordinates but the described method gives us different good coordinates.

Let h be a quasi-ordinary power series in good coordinates. Since $ND(h)$ is a monotone polygonal path we order the set $\{\gamma\}$ of compact 1-dimensional faces according to the slopes (which are rational d-tuples) of the edges using the reversed partial order defined before. In fact they are totally ordered and the highest one will be the z-highest. Assume that there are exactly r of them which are ordered as $\gamma_1, \ldots, \gamma_r$. For $q \in \{1, \ldots, r\}$, since γ_q is 1-dim, the weighted-homogeneous polynomial $f_{\gamma_q} = z^k \prod x_l^{a_l} (z^{n_1^q m_q} + \ldots + \alpha(x_1^{b_1^q} \ldots x_d^{b_d^q})^{m_q})$, for some non negative integers k, a_l and $(n_1^q, b_1^q, \ldots, b_d^q)$ relatively prime. Since γ_q is an edge of the Newton polytope then n_1^q is a positive integer, the d-tuple $(b_1^q, \ldots, b_d^q) \in \mathbb{N}^d$ is non-zero and $\alpha \in \mathbf{G}_{m,k}$. The polynomial $z^{-k} f_{\gamma_q}(\mathbf{1}, z^{1/n_1^q})$ of degree m_q in $k[z]$ can be factored. There are some positive integers $m_{q,j}, j = 1, \ldots, v(q)$, such that $f_{\gamma_q} = z^k \prod x_l^{a_l} f_{\boldsymbol{\lambda}_q}$ where

$$f_{\boldsymbol{\lambda}_q} := \prod_{j=1}^{v(q)} (f_{q,j})^{m_{q,j}}, \text{ where } f_{q,j} := z^{n_1^q} - \beta_j^q \mathbf{x}^{n_1^q \boldsymbol{\lambda}_q} = z^{n_1^q} - \beta_j^q x_1^{b_1^q} \ldots x_d^{b_d^q}.$$

We define the rational d-tuple $\boldsymbol{\lambda}_q := (\lambda_{q,1}, \ldots, \lambda_{q,d}) \in \mathbb{Q}^d$, where $\lambda_{q,l} = \frac{b_l^q}{n_1^q}$ for each $l = 1, \ldots, d$. We will say that $f_{\boldsymbol{\lambda}_q}$ is the weighted-homogeneous polynomial associated with the compact 1-dimensional face γ_q of $\Gamma(f)$.

REMARK 3.9. Let us see that we can find good coordinates which are not P-good coordinates for $f(x,z) = (z^2 - x^5)(z - x^2)(z^3 - x^2)$. The problem arises in the characteristic exponents which are integers. The definition-construction of P-good coordinates shows that there exists a change of coordinates of type $z_1 = z + a(\mathbf{x})$, $a(\mathbf{x}) \in k[[\mathbf{x}]]$, such that in the set $\Lambda_{ND}(h) := \{\boldsymbol{\lambda}_1, \ldots, \boldsymbol{\lambda}_r\}$, the greatest characteristic exponent $\boldsymbol{\lambda}_1$ is the only one which can be in $\mathbb{Z}^d$. The polynomial $f_{\boldsymbol{\lambda}_1}$ is called the *Newton initial form*; and in this case its Newton initial form $f_{\boldsymbol{\lambda}_1}$ cannot be a power of a linear form. In the above example doing the change

$z_1 = z + 2x^2$ we get a new Newton polytope which is a polygonal path with only two 1-dimensional faces. We observe two facts. The first one is that for each quasi-ordinary power series in good coordinates, the method described in [**23**, Lemma 3.16] provides P-good coordinates. The second one is that in any case, the characteristic exponents appearing in both Newton polyhedra which are smaller than the first integer characteristic exponent coincide in both polytopes. Our constructions work in P-good coordinates but proofs must be slightly modified.

PROPOSITION 3.10. *If h is in good coordinates then each $\boldsymbol{\lambda}_q$ belongs to $\Lambda_{CE}(f)$.*

PROOF. Since h is in good coordinates we can use the process given in [**36**] to compute the roots of f from the Newton polytope. In fact each root, different from zero, is found in only one of the edges of $ND(h)$. Assume $\gamma_1, \ldots, \gamma_r$ are the edges ordered as before. Then any pair $\zeta \in \gamma_q$ and $\zeta' \in \gamma_{q-1}$ gives $\boldsymbol{\lambda}_q$ as characteristic exponent, that is $\zeta - \zeta' = \mathbf{x}^{\boldsymbol{\lambda}_q}\epsilon(\mathbf{x})$, $\epsilon(\mathbf{0}) \neq 0$. If $\gamma_q = \gamma_1$, as h is in good coordinates, either z is a factor of h and $\zeta - 0 = \mathbf{x}^{\boldsymbol{\lambda}_1}$, $\zeta \in \gamma_1$ or f_{γ_1} is not a power of a linear form and taking two distinct roots in γ_1 gives $\boldsymbol{\lambda}_1$. □

Thus each root ζ_i of f is written in a unique way as

$$\zeta_i(\mathbf{x}) = \alpha_1^{(i)}\mathbf{x}^{\boldsymbol{\lambda}_{(i)}} + \sum_{\boldsymbol{\lambda}_{(i)} < \boldsymbol{\lambda}} \alpha_{\boldsymbol{\lambda}}^{(i)}\mathbf{x}^{\boldsymbol{\lambda}}, \ (\alpha_1^{(i)} \neq 0). \tag{3.4}$$

We define f_q to be the product of all roots ζ of f whose initial term gives the Newton polytope of $f_{\boldsymbol{\lambda}_q}$. This means that

$$f_q(\mathbf{x}, z) := \prod_{\text{roots:}\boldsymbol{\lambda}_{(i)} = \boldsymbol{\lambda}_q} (z - \zeta_i).$$

REMARK 3.11. Since all the conjugates under the Galois group of a root ζ_i have the same $\boldsymbol{\lambda}_{(i)}$, the irreducible component in $k[[\mathbf{x}]][z]$ of $f(\mathbf{x}, z)$ determined by ζ_i divides $f_q(\mathbf{x}, z)$. Thus $f_q \in k[[\mathbf{x}]][z]$ because it is a product of some irreducible components of $f(\mathbf{x}, z)$.

Thus we define the set $\Lambda_{ND}(h) := \{\boldsymbol{\lambda}_1, \ldots, \boldsymbol{\lambda}_r\}$ which is a subset of the characteristic exponents of h and which are ordered $\boldsymbol{\lambda}_r < \ldots < \boldsymbol{\lambda}_1$.

Let f_j^q be the element in $k[[\mathbf{x}]][z]$ defined by

$$f_j^q(\mathbf{x}, z) := \prod_{\text{roots:}\boldsymbol{\lambda}_{(i)} = \boldsymbol{\lambda}_q \text{ and } \beta^{(i)} = \beta_j^q} (z - \zeta_i).$$

The same ideas as in Remark 3.11 show that $f_j^q \in k[[\mathbf{x}]][z]$. Hence

$$f_q = \prod_{j=1}^{v(q)} f_j^q \quad \text{and} \quad f = \begin{cases} \displaystyle\prod_{q=1}^{r}\prod_{j=1}^{v(q)} f_j^q, & \text{if } I = I' \text{ and} \\ \displaystyle z\prod_{q=1}^{r}\prod_{j=1}^{v(q)} f_j^q, & \text{otherwise.} \end{cases}$$

DEFINITION 3.12. If h is in good coordinates then we call $\{f_j^q\}_{q,j}$ the *Newton components of* f; each f_j^q defines a quasi-ordinary power series.

3. Dual decomposition

Assume a pair (h, ω) satisfying the support condition 1.3 is given, where h is a quasi-ordinary power series in good coordinates as in (3.2) and $\omega = \prod_{j=1}^{d} x_j^{\nu_j - 1} dx_1 \wedge \ldots \wedge dx_d \wedge dz$, $\nu_j \geq 1$. The compact faces of $\Gamma(f)$ (or $\Gamma(h)$) are 1-dim edges $\gamma_1, \ldots, \gamma_r$ with their corresponding vertices $\tau_0, \tau_1, \ldots, \tau_r$.

REMARK 3.13. The main point in the Newton polytope is that its set of compact faces (and then its vertices) is totally ordered. We recall that τ_r has the highest z-coordinate.

Let $e_1, \ldots, e_{d+1}$ denote the canonical basis of the dual space $(\mathbb{R}^{d+1})^*$ where we choose coordinates $(v_1, \ldots, v_d, v_{d+1})$. The fan $\Sigma(\Gamma(h)) \subset (\mathbb{R}^{d+1})^*$ is obtained subdividing the cone $(\mathbb{R}_+^{d+1})^*$ with linear hyperplanes $l^q : \eta_q = 0$ where

$$\eta_q := \sum_{l=1}^{d} b_l^q v_l - n_1^q v_{d+1} = n_1^q \left(\sum_{l=1}^{d} \lambda_{q,l} v_l - v_{d+1} \right), \quad q = 1, \ldots, r. \tag{3.5}$$

Let $\Delta_{\gamma_q} = \{ \mathbf{k} \in \mathbb{R}_+^{d+1} \mid \mathbf{k} \cdot (b_1^q, \ldots, b_d^q, -n_1^q) = 0 \}$ be its dual cone.

LEMMA 3.14. *Δ_{γ_q} is the strictly positive simplicial cone $\Delta_{\gamma_q} = \{\lambda_1 \mathbf{w}_1^q + \ldots + \lambda_d \mathbf{w}_d^q,\ \lambda_i \in \mathbb{R}_+\}$, where $\mathbf{w}_l^q := \frac{1}{c_l^q}(n_1^q e_l + b_l^q e_{d+1}) \in \mathbb{N}^{d+1}$, $c_l^q := \gcd(n_1^q, b_l^q)$ with $l = 1 \ldots, d$.*

The linearly independent vectors $\{\mathbf{w}_l^q\}_{l=1,..d}$ are primitive, so the proof is clear. Let G_q be the fundamental set of Δ_{γ_q}, then (2.1) implies

$$\Phi_{\Delta_{\gamma_q}}(\mathbf{y}) := \Phi_{\Delta_{\gamma_q} \cap \mathbb{P}^{d+1}}(\mathbf{y}) = \frac{\sum_{\beta \in G_q} \mathbf{y}^{\beta}}{\prod_{l=1}^{d} (1 - \mathbf{y}^{\mathbf{w}_l^q})}. \tag{3.6}$$

If $p = (p_1, \ldots, p_d, p_{d+1})$ is an element in the closure of γ_q then (2.5) gives

$$S_{\Delta_{\gamma_q}}(h, \omega, T) = \Phi_{\Delta_{\gamma_q}}(\mathbb{L}^{-\nu_1} T^{p_1}, \ldots, \mathbb{L}^{-\nu_{d+1}} T^{p_{d+1}}). \tag{3.7}$$

The edge γ_q is defined by the affine equations $n_1^q x_l + b_l^q z = M_l^q, \quad l = 1, \ldots, d$, for some positive integers M_l^q.

LEMMA 3.15. *$S_{\Delta_{\gamma_q}}(h, \omega, T) \in \mathbb{Z}[\mathbb{L}, \mathbb{L}^{-1}, (1 - \mathbb{L}^{-(\nu_l p_l^q + \bar{b}_l^q)} T^{\frac{M_l^q}{c_l^q}})^{-1}][T]$, for every $1 \leq l \leq d$, and*

$$J_{\gamma_q}(h, \omega, s) = \frac{(n_1^q)^{d-1}}{\tilde{T_1}^q \ldots \tilde{T_d}^q},$$

where $\bar{b}_l^q := \frac{b_l^q}{c_l^q}$, $p_l^q := \frac{n_1^q}{c_l^q}$, $\tilde{T_l}^q := (M_l^q s + \nu_l n_1^q + b_l^q)$, $l = 1, \ldots, d$ and $q = 1, \ldots, r$.

PROOF. Since $m_h(\mathbf{w}_l^q) = \mathbf{w}_l^q \cdot \mathbf{a}$, where $\mathbf{a} \in \gamma_q$, then $m_h(\mathbf{w}_l^q) = \frac{M_l^q}{c_l^q}$. On the other hand, $\sigma_w(\mathbf{w}_l^q) = \mathbf{w}_l^q \cdot (\nu_1, \ldots, \nu_d, 1) = \frac{1}{c_l^q}(\nu_l n_1^q + b_l^q)$. Let us compute $J_{\gamma_q}(h, \omega, s)$. Because of $\gcd(b_1^q, \ldots, b_d^q, -n_1^q) = 1$ we choose $\beta := (a_1, \ldots, a_l, u)$ in order to obtain $(b_1^q, \ldots, b_d^q, -n_1^q) \cdot \beta = 1$. Computing $\mathrm{mult}(\Delta_{\gamma_q}) = |\det(\mathbf{w}_1^q, \ldots, \mathbf{w}_d^q, \beta)|$

by induction one can prove that

$$\det(\mathbf{w}_1^q,\ldots,\mathbf{w}_d^q,\beta) = -(n_1)^{d-1}\frac{(b_1^q,\ldots,b_d^q,-n_1^q)\cdot\beta}{\prod_{l=1}^d c_l^q} = -\frac{(n_1^q)^{d-1}}{\prod_{l=1}^{d} c_l^q}.$$

The result follows from (2.7). □

Consider next the first vertex τ_0 with coordinates $(\mathbf{a},d_0)$ in $\Gamma(h)$. The rational simplicial cone Δ_{τ_0} is the $(d+1)$-dimensional cone: $\mathbb{R}_+\mathbf{w}_1^1+\cdots+\mathbb{R}_+\mathbf{w}_d^1+\mathbb{R}_+e_{d+1}$. In a similar manner as above if G_{τ_0} is the fundamental set of Δ_{τ_0} then

$$\Phi_{\Delta_{\tau_0}}(\mathbf{y}) = \frac{\left(\sum_{\beta\in G_{\tau_0}}\mathbf{y}^{\beta}\right)}{(1-y_{d+1})\prod_{l=1}^d\left(1-\mathbf{y}^{\mathbf{w}_l^1}\right)},$$

and

(3.8) $$S_{\Delta_{\tau_0}}(h,\omega,T) = \Phi_{\Delta_{\tau_0}}(\mathbb{L}^{-\nu_1}T^{a_1},\ldots,\mathbb{L}^{-\nu_d}T^{a_d},\mathbb{L}^{-1}T^{d_0}).$$

Recall that $I = I'$ if and only if $d_0 = 0$; otherwise $d_0 = 1$.

LEMMA 3.16.

$$S_{\Delta_{\tau_0}}(h,\omega,T) \in \mathbb{Z}[\mathbb{L},\mathbb{L}^{-1},(1-\mathbb{L}^{-1}T)^{-\varepsilon},(1-\mathbb{L}^{-(\nu_l p_1^1+\bar{b}_l^1)}T^{\frac{M_l^\tau}{c_l^1}})^{-1}][T]$$

for every $1\le l\le d$, *and*

$$J_{\tau_0}(h,\omega,s) = \frac{(n_1^1)^d}{(s+1)^{\varepsilon}\tilde{T}_1^1\ldots\tilde{T}_d^1}, \quad \textit{see Lemma 3.15},$$

where $\varepsilon = 0$ *or* 1 *when* $I = I'$ *or not.*

PROOF. If $I=I'$, $m_h(e_{d+1}) = 0$ and $\sigma_\omega(e_{d+1}) = 1$. Furthermore, $\mathrm{mult}(\Delta_{\tau_0}) = |\det(\mathbf{w}_1^1,\ldots,\mathbf{w}_d^1,e_{d+1})| = \dfrac{(n_1^1)^d}{\prod_{l=1}^{d} c_l^1}$, and we are done. The other case is similar. □

Finally consider any other vertex τ_q of $\Gamma(h)$, different from τ_0, which is the intersection of the edges γ_q and γ_{q+1}, $q=1,\ldots,r$. If $q=r$ then γ_{r+1} is a (non-compact) line parallel to z-axis.

The dual cone Δ_{τ_q} is $\sum_{i=1}^d(\mathbb{R}_+\mathbf{w}_i^q+\mathbb{R}_+\mathbf{w}_i^{q+1})$, where if $q=r$ then $\mathbf{w}_l^{r+1}:=e_l$, for $l=1,\ldots,d$. For any $q=1,\ldots,r$, consider the strictly positive simplicial cone

$$\Delta_q^c := \sum_{l=1}^d \mathbb{R}_+\mathbf{w}_l^q + \mathbb{R}_+e_{d+1}.$$

For instance Δ_{r+1}^c is the whole positive cone $\mathbb{P}^{d+1}$. Moreover its multiplicity satisfies

(3.9) $$\mathrm{mult}(\Delta_q^c) = |\det(\mathbf{w}_1^q,\ldots,\mathbf{w}_d^q,e_{d+1})| = \frac{(n_1^q)^d}{\prod_{l=1}^d c_l^q}.$$

The following relation holds for the indicator functions of these cones:

(3.10) $$[\Delta_{\tau_q}] - [\Delta_{q+1}^c] + [\Delta_{\gamma_q}] + [\Delta_q^c] = 0, \qquad q=1\ldots,r,$$

which implies the same relation among the corresponding generating functions. Since Δ_q^c and Δ_{γ_q} are simplicial cones,

$$\Phi_{\Delta_{\tau_q}}(\mathbf{y}) = \Phi_{\Delta_{q+1}^c}(\mathbf{y}) - \left(\Phi_{\Delta_{\gamma_q}}(\mathbf{y}) + \Phi_{\Delta_q^c}(\mathbf{y})\right). \tag{3.11}$$

If G_q^c is the fundamental set of Δ_q^c then

$$\Phi_{\Delta_{\tau_q}}(\mathbf{y}) = \frac{\left(\sum_{\beta\in G_{q+1}^c}\mathbf{y}^{\beta}\right)}{(1-y_{d+1})\prod_{l=1}^d\left(1-\mathbf{y}^{\mathbf{w}_l^{q+1}}\right)} - \left(\frac{\left(\sum_{\beta\in G_q}\mathbf{y}^{\beta}\right)}{\prod_{l=1}^d\left(1-\mathbf{y}^{\mathbf{w}_l^q}\right)} + \frac{\left(\sum_{\beta\in G_q^c}\mathbf{y}^{\beta}\right)}{(1-y_{d+1})\prod_{l=1}^d\left(1-\mathbf{y}^{\mathbf{w}_l^q}\right)}\right).$$

If $\tau_q \in \Gamma(h)$ has z-height d_q then

$$S_{\Delta_{\tau_q}}(h,\omega,T) = \Phi_{\Delta_{\tau_q}}(\mathbb{L}^{-\nu_1}T^{a_1},\ldots,\mathbb{L}^{-\nu_d}T^{a_d},\mathbb{L}^{-1}T^{d_q}). \tag{3.12}$$

Hence since $m_h(e_j) = N_j$ and $\sigma_\omega(e_j) = \nu_j$ then

$$S_{\Delta_{\tau_r}}(h,\omega,T) \in \mathbb{Z}\left[\mathbb{L}^{\pm1}, \frac{1}{1-\mathbb{L}^{-\nu_j}T^{N_j}}, \frac{1}{1-\mathbb{L}^{-(\nu_l p_l^r+\bar{b}_l^r)}T^{\frac{M_l^r}{c_l^r}}}\right][T], \tag{3.13}$$

with $j, l \in \{1,\ldots,d\}$. In the same way, for each $q \in \{1,\ldots,r-1\}$,

$$S_{\Delta_{\tau_q}}(h,\omega,T) \in \mathbb{Z}\left[\mathbb{L}^{\pm1}, \frac{1}{1-\mathbb{L}^{-(\nu_l p_l^q+\bar{b}_l^q)}T^{\frac{M_l^q}{c_l^q}}}, \frac{1}{1-\mathbb{L}^{-(\nu_l p_l^{q+1}+\bar{b}_l^{q+1})}T^{\frac{M_l^{q+1}}{c_l^{q+1}}}}\right][T], \tag{3.14}$$

with $l = 1,\ldots,d$.

To compute the local topological zeta function more explicitly we argue as follows. Recall that $J_{\tau_q}(h,\omega,s)$ can be computed using an adequate simplicial decomposition of Δ_{τ_q}. We use the decomposition from (3.11). Then

$$J_{\tau_q}(h,\omega,s) = \chi_{\text{top}}\left((\mathbb{L}-1)^{d+1}\left(\Phi_{\Delta_{q+1}^c}(\mathbf{a}) - \left(\Phi_{\Delta_{\gamma_q}}(\mathbf{a}) + \Phi_{\Delta_q^c}(\mathbf{a})\right)\right)\right),$$

where $\mathbf{a} := (\mathbb{L}^{-(\nu_1+a_1 s)},\ldots,\mathbb{L}^{-(\nu_d+a_d s)},\mathbb{L}^{-(1+d_q s)})$. Since $\Phi_{\Delta_{\gamma_q}}(\mathbf{a})$ has at most d poles then

$$J_{\tau_q}(h,\omega,s) = \chi_{\text{top}}\left((\mathbb{L}-1)^{d+1}\left(\Phi_{\Delta_{q+1}^c}(\mathbf{a}) - \Phi_{\Delta_q^c}(\mathbf{a})\right)\right).$$

Moreover we know from equation (3.9) that $\operatorname{mult}(\Delta_q^c) = \dfrac{(n_1^q)^d}{\prod_{l=1}^d c_l^q}$.

COROLLARY 3.17. *For each $q \in \{1,\ldots,r-1\}$, then*

$$J_{\tau_q}(h,\omega,s) = \frac{1}{(1+d_q s)}\left(\frac{(n_1^{q+1})^d}{\tilde{T_1}^{q+1}\ldots\tilde{T_d}^{q+1}} - \frac{(n_1^q)^d}{\tilde{T_1}^q\ldots\tilde{T_d}^q}\right),$$

and if $q = r$ then

$$J_{\tau_r}(h,\omega,s) = \frac{1}{(1+d_r s)}\left(\frac{1}{T_1\ldots T_d} - \frac{(n_1^r)^d}{\tilde{T_1}^r\ldots\tilde{T_d}^r}\right),$$

where $T_i := N_i s + \nu_i$, $\tilde{T_l^r}$ have been defined in Lemma 3.15, and $i, l \in \{1,\ldots,d\}$.

REMARK 3.18. It follows from the above discussion, Definition 2.3 and Remark 3.19 that

$$Z_{DL}^{A}(h,\omega,T)=\sum_{j=0}^{r}L_{\tau_j}^{A}(h)S_{\Delta_{\tau_j}}(h,\omega,T)+\sum_{q=1}^{r}L_{\gamma_q}^{A}(h)S_{\Delta_{\gamma_q}}(h,\omega,T)= \tag{3.15}$$

$$=\mathbb{L}^{-(d+1)}(\mathbb{L}-1)^{d+1}\left(\sum_{j=0}^{r}S_{\Delta_{\tau_j}}(h,\omega,T)+\sum_{q=1}^{r}\left(1-\frac{v(q)}{\mathbb{L}-1}\right)S_{\Delta_{\gamma_q}}(h,\omega,T)\right),$$

$$Z_{\mathrm{top},0}^{A}(h,\omega,s)-\sum_{j=0}^{r}J_{\tau_j}(h,\omega,s)-\sum_{q=1}^{r}v(q)J_{\gamma_q}(h,\omega,s). \tag{3.16}$$

After Lemma 2.1, each $(1-\mathbb{L}^{-1}T^{d_q})$ or $(1+d_q s)$ appearing in (3.12) and Corollary 3.17 does not contribute to the denominator of the A-part of the corresponding zeta functions of (h,ω).

4. Newton map associated with a Newton component

We have already discussed the dual decomposition associated with the Newton polyhedra of quasi-ordinary power series in good coordinates. In general these power series are degenerate with respect to their Newton polyhedron. In this section we describe how to *improve* a quasi-ordinary power series in terms of the complexity of the series.

Each compact 1-dimensional face γ_q of $\Gamma(f)$ corresponds to a polynomial

$$f_{\boldsymbol{\lambda}_q}=\prod_{j=1}^{v(q)}(z^{n_1^q}-\beta_j^q\mathbf{x}^{n_1^q\boldsymbol{\lambda}_q})^{m_{q,j}}=\prod_{j=1}^{v(q)}(z^{n_1^q}-\beta_j^q x_1^{b_1^q}\dots x_d^{b_d^q})^{m_{q,j}}.$$

Fix $j\in\{1,\dots,v(q)\}$, the corresponding polynomial $f_{q,j}=z^{n_1^q}-\beta_j^q\mathbf{x}^{n_1^q\boldsymbol{\lambda}_q}$ and the series f_j^q, see Definition 3.12. We denote by $\mathbf{ev}(f_{q,j})=:r\le d$, the number of essential variables of the irreducible factor $f_{q,j}$. For the sake of simplicity we assume that $x_1,\dots,x_r$ are the essential variables of $f_{q,j}$. Fix $\pi_j^q:\mathbb{A}_k^r\to V_j^q\subset\mathbb{A}_k^{r+1}$ a morphism (a parametrization) of the irreducible variety defined by $f_{q,j}$:

$$(s_1,\dots,s_r)\mapsto(s_1^{p_1^q},\dots,s_r^{p_r^q},\alpha_j^q s_1^{\bar b_1^q}\dots s_r^{\bar b_d^r}),$$

where $(\alpha_j^q)^{n_1^q}=\beta_j^q$.

REMARK 3.19. The morphism π_j^q is surjective but it is not injective. The variety V_j^q is the quotient of $\mathbb{A}_k^r$ under the action of a finite abelian group G. By [**35**, Lemma 5.1] (see also [**18**]) in $K_0(\mathrm{Var}_k)$ the following classes are the same object $[V_j^q]=[\mathbb{A}_k^r/G]=\mathbb{L}^r$. Later on we will consider the subvariety $(V_j^q\times\mathbb{A}_k^{d-r})\cap\mathbf{G}_{m,k}^{d+1}$ on the torus; its class in $K_0(\mathrm{Var}_k)$ is $(\mathbb{L}-1)^d$. Since $f_{\boldsymbol{\lambda}_q}$ has $v(q)$ irreducible components, $[N_{\gamma_q}]=v(q)(\mathbb{L}-1)^d$.

DEFINITION 3.20. Let $X=\operatorname{Spec}k[[\mathbf{x},z]],Y=\operatorname{Spec}k[[\mathbf{y},z_1]]$. The *Newton map* associated with the Newton component f_j^q (or with the polynomial $f_{q,j}$) of f is the morphism $\pi_{q,j}:Y\to X$, $\pi_{q,j}(\mathbf{y},z_1):=(\mathbf{x},z)$ defined by the equations

$$x_l=y_l^{p_l^q},\ l=1,\dots,d\quad\text{and}\quad z=(z_1+\alpha_j^q)\prod_{l=1}^{d}y_l^{\bar b_l^q}.$$

The integers $\bar{b}_l^q$, p_l^q were defined in Lemma 3.14 and if $l = r+1, \ldots, d$, then $\bar{b}_l^q := 0$ and $p_l^q := 1$.

LEMMA 3.21. *Let h be a quasi-ordinary power series in good coordinates. The pull-back $\bar{h}_{q,j}$ of h under the Newton map $\pi_{q,j}$ associated with a Newton component f_j^q of f defines a quasi-ordinary power series at the root 0. Moreover under any Newton map $\mathbf{ev}(h)$ does not increase. More precisely the set of essential variables of $\bar{h}_{q,j}$ is contained in the one of h (identifying x_l and y_l).*

PROOF. From the well-known properties of the discriminant the following identities will give the proof of the lemma.

$$D_{z_1}(h \circ \pi_{q,j}) = y_1^{m_1} y_2^{m_2} \ldots y_d^{m_d} (D_z(h) \circ \pi_{q,j}) = y_1^{r_1} y_2^{r_2} \ldots y_d^{r_d} \varepsilon(\mathbf{y}),$$

where $\varepsilon(0) \neq 0, m_i \geq 0, r_i \geq 0$. The conditions on $\mathbf{ev}(h)$ and the essential variables are also clear. □

Let f_k^i be a Newton component of f, hence it defines a quasi-ordinary power series. The previous lemma shows that the pull-back $f_k^i \circ \pi_{q,j}$ can be decomposed as $y_1^{a_1} y_2^{a_2} \ldots y_d^{a_d} \bar{f}_k^i$, for some power series $\bar{f}_k^i \in k[[\mathbf{y}, z_1]]$.

LEMMA 3.22.

(1) *If $\boldsymbol{\lambda}_{(i)} = \boldsymbol{\lambda}_q$ and $\beta_k^i = \beta_j^q$ then $\bar{f}_k^i(\mathbf{0}) = 0$, therefore $\bar{f}_k^i$ defines a quasi-ordinary power series.*
(2) *Otherwise, $\bar{f}_k^i(\mathbf{0}) \neq 0$, and $\bar{f}_k^i$ is a unit in $k[[\mathbf{y}, z_1]]$.*

PROOF. The proof of the lemma will follow from the following description. Take one of the roots ζ of f different from $z = 0$. Since we are in a good system of coordinates, ζ is written as, see identity (3.4),

$$\zeta = \alpha_1^{(i)} \mathbf{x}^{\boldsymbol{\lambda}_{(i)}} + \sum_{\boldsymbol{\lambda}_{(i)} < \boldsymbol{\lambda}} \alpha_{\boldsymbol{\lambda}}^{(i)} \mathbf{x}^{\boldsymbol{\lambda}}.$$

The pull-back of ζ under the Newton map $\pi_{q,j}$ is as follows.

(1) If $\boldsymbol{\lambda}_{(i)} = \boldsymbol{\lambda}_q$ and $(\alpha_1^{(i)})^{n_1^{(i)}} = \beta_j^q$ then

$$(z - \zeta) \circ \pi_{q,j}(\mathbf{y}, z_1) = \mathbf{y}^{\boldsymbol{\lambda}_q}(z_1 - \sum_{\boldsymbol{\lambda}_q < \boldsymbol{\lambda}} \alpha_{\boldsymbol{\lambda}}^{(i)} \mathbf{y}^{\boldsymbol{\lambda} - \boldsymbol{\lambda}_q}).$$

(2) If $\boldsymbol{\lambda}_{(i)} = \boldsymbol{\lambda}_q$ but $(\alpha_1^{(i)})^{n_1^{(i)}} \neq (\alpha_j^q)^{n_1^q} = \beta_j^q$ then

$$(z - \zeta) \circ \pi_{q,j}(\mathbf{y}, z_1) = \mathbf{y}^{\boldsymbol{\lambda}_q}(z_1 + \alpha_j^q - \alpha_1^{(i)} - \sum_{\boldsymbol{\lambda}_q < \boldsymbol{\lambda}} \alpha_{\boldsymbol{\lambda}}^{(i)} \mathbf{y}^{\boldsymbol{\lambda} - \boldsymbol{\lambda}_q}).$$

(3) If $\boldsymbol{\lambda}_{(i)} < \boldsymbol{\lambda}_q$ then

$$(z - \zeta) \circ \pi_{q,j}(\mathbf{y}, z_1) = \mathbf{y}^{\boldsymbol{\lambda}_{(i)}}((z_1 + \alpha_j^q)\mathbf{y}^{\boldsymbol{\lambda}_q - \boldsymbol{\lambda}_{(i)}} - \alpha_1^{(i)} - \sum_{\boldsymbol{\lambda}_{(i)} < \boldsymbol{\lambda}} \alpha_{\boldsymbol{\lambda}}^{(i)} \mathbf{y}^{\boldsymbol{\lambda} - \boldsymbol{\lambda}_{(i)}}).$$

(4) If $\boldsymbol{\lambda}_q < \boldsymbol{\lambda}_{(i)}$ then

$$(z - \zeta) \circ \pi_{q,j}(\mathbf{y}, z_1) = \mathbf{y}^{\boldsymbol{\lambda}_q}(z_1 + \alpha_j^q - \alpha_1^{(i)} \mathbf{y}^{\boldsymbol{\lambda}_{(i)} - \boldsymbol{\lambda}_q} - \sum_{\boldsymbol{\lambda}_{(i)} < \boldsymbol{\lambda}} \alpha_{\boldsymbol{\lambda}}^{(i)} \mathbf{y}^{\boldsymbol{\lambda} - \boldsymbol{\lambda}_q}).$$

□

The pull-back $\bar{h}_{q,j}$ of h is $\prod_{l=1}^{d} y_l^{N_l} \bar{f}_{q,j}(\mathbf{y}, z_1) w(\mathbf{y}, z_1)$ where $w(\mathbf{0}) \neq 0$. Moreover the characteristic exponents of the quasi-ordinary power series $\bar{f}_{q,j}(\mathbf{y}, z_1)$ are easily deduced from the characteristic exponents of f. In general the new quasi-ordinary power series $\bar{h}_{q,j}$ is not given in a good system of coordinates. Nevertheless the change of coordinates we need to put it in good coordinates is described in Remark 3.8.

REMARK 3.23. If β_j^q is a simple root of the polynomial $f_{\boldsymbol{\lambda}_{\kappa_q}}(\mathbf{1}, z)$ then $\bar{f}_j^q(\mathbf{0}, z_1) = z_1 + \cdots$. Using the Implicit Function Theorem the root of $\bar{f}_j^q$ is a series in $k[[\mathbf{y}]]$. As in the Puiseux Theorem ($d = 1$), all roots of f can be found by using a sequence of Newton maps, (see [**36**, Theorem 2] for details)

If f (or h) has non-degenerate Newton principal part then all roots of all $f_{\boldsymbol{\lambda}_q}(\mathbf{1}, z)$ are simple roots. Otherwise f has degenerate principal part and following the proof of [**36**, Theorem 2] one can use successive Newton maps and maps of type $z_1 = z - m(\mathbf{x})$ to get again a good system of coordinates and reach a point at which all pull-backs of f have non-degenerate principal part.

DEFINITION 3.24. We define the *depth* of a quasi-ordinary power series h, and denote it by depth(h), as follows. First we put h in P-good coordinates; assume h is represented using its Newton components as follows:

$$h = x_1^{N_1} x_2^{N_2} \dots x_d^{N_d} z^{\varepsilon} \prod_{q=1}^{r} \prod_{j=1}^{u(q)} f_j^q(\mathbf{x}, z) u(\mathbf{x}, z), \qquad \text{with } u(\mathbf{0}, 0) \neq 0.$$

If $h = x_1^{N_1} x_2^{N_2} \dots x_d^{N_d} z^{\varepsilon} u(\mathbf{x}, z)$, $\varepsilon = 0, 1$, then we will say that depth$(h) := 0$. Otherwise

$$\operatorname{depth}(h) := \max\{\operatorname{depth}(\bar{h}_{q,j})\} + 1,$$

where the maximum is taken over all pull-backs $\bar{h}_{q,j}$ under the composition of the Newton maps associated with Newton components f_j^q of f and a coordinate change to obtain P-good coordinates. If h appears in good coordinates and a choice η of good coordinates has been given after each Newton mapping, then we define in the same way depth$_\eta(h)$.

REMARK 3.25. The above discussion shows that depth$(h) \in \mathbb{N}$ and decreases under the Newton maps. Moreover h has non-degenerate Newton principal part and $h \neq x_1^{N_1} x_2^{N_2} \dots x_d^{N_d} z^{\varepsilon} u(\mathbf{x}, z)$ (u unity) if and only if depth$(h) = 1$. The finiteness of the invariant depth(h) will be used in the proof of the main result. It is easily seen that for any recursive choice η of good coordinates, depth$_\eta(h) \leq$ depth(h). Note that the components which correspond to edges appearing after an edge with integer multi-slope need less Newton mappings to be *improved.*

NEWTON TREES 3.26. We can keep all the information about the Newton process in a tree, the same way that Eisenbud and Neumann diagrams do for curves. Assume we have chosen a system of coordinates such that the Newton polygon of h is a polygonal path. In our tree, each compact face will be represented by a vertex, and two vertices are connected by an edge (a vertical edge) if and only if the two faces intersect. The non compact faces are represented by arrows connected to the face they intersect. The arrow $\mathcal{F}_{0,\dots,0}$ representing $x_i = N_i, i = 1, ..., d$ is decorated by $(N_1, ..., N_d)$, and the arrow $\mathcal{F}_0$, representing the hyperplane $z = 0$ or $z = 1$, is decorated by 0 or 1. If the polygonal line consists in one point $(N_1, N_2, ..., N_d, N)$, its associated graph is a vertical edge, with two arrows at the ends which represent the above non compact faces: the arrow at the top is decorated with $(N_1, N_2, \cdots, N_d)$ and the arrow at the bottom is decorated with N.

More generally, each compact face γ_q is defined by the affine equations $n_1^q x_l + b_l^q z = M_l^q, \quad l = 1, \ldots, d$, for some positive integers M_l^q, see (3.7). If γ_q is represented by the vertex v_q we decorate v_q with $(M_1^q, \ldots, M_d^q)$ and the end of the edge under the vertex v_q is decorated with n_1^q. The *numerical datum* of the vertex v_q is $((M_1^q, \ldots, M_d^q), n_1^q)$.

Now if f is non-degenerate with respect to γ_q, we attached to v_q as many non vertical edges ending with arrows as the number of integer points on γ_q minus one (recall that this is $v(q)$). If not, we consider a Newton map attached to γ_q and one of its factors. We consider the pull back of h under this map, and put it in coordinates such that the Newton polygon is a polygonal path. We consider the part of the tree corresponding to the new Newton polygon and we delete its arrow $\mathcal{F}_{0,..,0}$ and attached the edge to the vertex v_q, so that the edge is not vertical. Because after a finite number of steps we have something non-degenerate, at the end we have a tree whose edges are attached to vertices or arrows. We can make this diagram minimal by first erasing the edges at the bottom of the tree ending with an arrow decorated with a 0 (we also erase this arrow) and decorated with 1, and then the vertex on the other end of the edge if it is connected only to two other vertices. Diagrams in good and P-good coordinates have the same vertices with the same numerical data, see [**4**].

EXAMPLE 3.27. **Newton trees for quasi-ordinary power series**. It is easy to check that the quasi-ordinary power series

$$h := (((z^2 - x_1^3)^2 - 4x_1^7)^2 - x_1^{20}x_2^3)((z^2 - x_1^4x_2^3)^2 - x_1^{11}x_2^7) \in \mathbb{C}[[x_1, x_2]][z]$$

is in good coordinates. Its Newton principal part is given by $(z^2 - x_1^3)^4(z^2 - x_1^4x_2^3)^2$ which is degenerate. Its Newton polygon has two compact 1-dimensional faces γ_1 (corresponding to $z^2 - x_1^3$) and γ_2 (corresponding to $z^2 - x_1^4x_2^3$). Below we give the affine equations defining the edges of the Newton polygon:

$$\gamma_1 : 2x_1 + 3z = 36, x_2 = 0,$$

$$\gamma_2 : 2x_1 + 4z = 40, \ 2x_2 + 3z = 12.$$

We have two Newton components

$$f_1^1 := ((z^2 - x_1^3)^2 - 4x_1^7)^2 - x_1^{20}x_2^3 \text{ and } f_1^2 := (z^2 - x_1^4x_2^3)^2 - x_1^{11}x_2^7.$$

The invariants for such faces are $n_1^1 = 2$, $\boldsymbol{\lambda}_1 = (3/2, 0)$, $(b_1^1, b_2^1) = (3, 0)$ and $n_1^2 = 2$, $\boldsymbol{\lambda}_2 = (2, 3/2)$, $(b_1^2, b_2^2) = (4, 3)$. On the other hand $(M_1^1, M_2^1) = (36, 0)$ and $(M_1^2, M_2^2) = (40, 12)$. These are the numerical data $\{((M_1^q, M_2^q), n_1^q)\}_{q=1}^2$ which appear in the (LHS) of the corresponding splice diagram in Figure 1.

Since h is degenerate and we have to start the process with two Newton components we must do two Newton maps. The map $\pi_{1,1}$ (corresponding to f_1^1) is given by

$$x_1 = y_1^2, x_2 = y_2, z = (z_1 + 1)y_1^3$$

and the map $\pi_{2,1}$ (corresponding to f_1^2) is given by

$$x_1 = \bar{y}_1, x_2 = \bar{y}_2^2, z = (\bar{z}_1 + 1)\bar{y}_1^2\bar{y}_2^3.$$

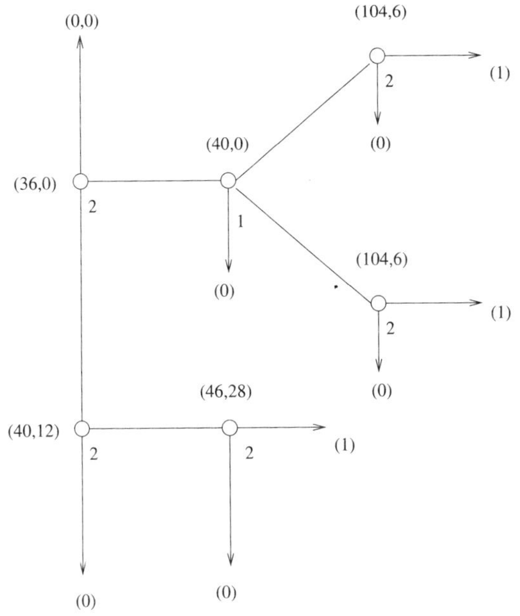

FIGURE 1.

In Figure 1, for each Newton map we have an edge from each corresponding vertex. The pull-backs of h under such maps, up to a unit, are given by

$\bar{h}_{1,1} := y_1^{36}((((z_1+1)^2-1)^2-4y_1^2)^2-y_1^{16}y_2^3)$ and $\bar{h}_{2,1} := \bar{y}_1^{20}\bar{y}_2^{12}(((\bar{z}_1+1)^2-1)^2-\bar{y}_1^3\bar{y}_2^2)$.

For $\bar{h}_{2,1}$, after a change of local coordinates only involving $\bar{z}_1$, the new quasi-ordinary power series is given $\bar{y}_1^{20}\bar{y}_2^{12}(\bar{z}_1^2-\bar{y}_1^3\bar{y}_2^2)$. The equations for the unique face γ are $\gamma : 2x_1 + 3z = 46, 2x_2 + 2z = 28$. Here we are in a non-degenerate case and we get the decorations $((46,28),2)$ in the diagram of the Figure 1.

The quasi-ordinary power series $\bar{h}_{1,1}$ has one compact 1-dimensional face given by $y_1^{36}((2z_1+2y_1)(2z_1-2y_1))^2$ with two Newton components:

$$\bar{h}_{1,1} = y_1^{36}(((z_1^2+2z_1+2y_1)(z_1^2+2z_1-2y_1))^2 - y_1^{16}y_2^3);$$

the decoration $((40,0),1)$ comes from this edge. We deal with one of the Newton components since the other one behaves in a similar way. The Newton map π corresponding to it is given by

$$y_1 = w_1,\, y_2 = w_2,\, z_1 = (z_2+1)w_1.$$

The pull back $\tilde{h}$ of $\bar{h}_{1,1}$ under π is

$$\tilde{h} = w_1^{40}(((w_1z_2^2+2w_1z_2+2z_2+w_1+4)(w_1z_2^2+2w_1z_2+2z_2+w_1))^2 - w_1^{12}w_2^3)$$

and after a local change of the z_2 coordinate, we can write down the equation of $\tilde{h}$ as $w_1^{40}(\tilde{z}_2^2 - w_1^{12}w_2^3)$ which is non-degenerate. This provides the numerical data $((104,6);2)$ in the diagram of the Figure 1.

We can do the diagram minimal erasing the edge with a (1) down the vertex corresponding to $(40,0)$.

DEFINITION 3.28. Let h be a quasi-ordinary power series with a recursive choice η of good coordinates. As usual suppose $h = x_1^{N_1} x_2^{N_2} \dots x_d^{N_d} f(\mathbf{x}, z) u(\mathbf{x}, z)$ and assume f is represented using its Newton components as: $f = z^{\varepsilon} \prod_{q=1}^{r} \prod_{j=1}^{v(q)} f_j^q$, $\varepsilon = 0$ or 1. Let ω be a differential form such that the pair (h, ω) satisfies the support condition (1.3). According to Lemmas 3.15, 3.16, and equations (3.13) and (3.14) we consider a set $CP_\eta(h, \omega)$ of *candidate poles* for (h, ω) recursively as follows:

$$CP_\eta(h,\omega) := \{(1,1)^{\varepsilon}, (N_i, \nu_i)\}_{i=1}^{d} \cup \bigcup_{q=1}^{r} \left\{ \left(\frac{M_l^q}{c_l^q}, \nu_l p_l^q + \bar{b}_l^q \right) \right\}_{l=1}^{d} \cup \bigcup CP_\eta(\bar{h}_{q,j}, \bar{\omega}_{q,j}),$$

where the last union runs over all pull-backs $(\bar{h}_{q,j}, \bar{\omega}_{q,j})$ of (h, ω) under the Newton maps associated with Newton components f_j^q of f, $q = 1, \dots, r$ and $j = 1, \dots, v(q)$. We will drop η if P-good coordinates have been chosen.

EXAMPLE 3.29. We continue with the quasi-ordinary power series h from Example 3.27. We also consider the differential form $\omega = dx_1 \wedge dx_2 \wedge dz$ which satisfies the support condition. Here $\varepsilon = 0$, $N_i = 0$ and $\nu_i = 1$ for $1 \le i \le 2$.

Below we give the affine equations defining the edges of the Newton polygon and the equations of the candidate pole:

$$\gamma_1 : 2x_1 + 3z = 36, x_2 = 0, \ (36s + 5);$$

$$\gamma_2 : 2x_1 + 4z = 40, \ 2x_2 + 3z = 12, \ (20s + 3) \ (12s + 5).$$

This data can be also encoded in the corresponding splice diagram (see Figure 2).

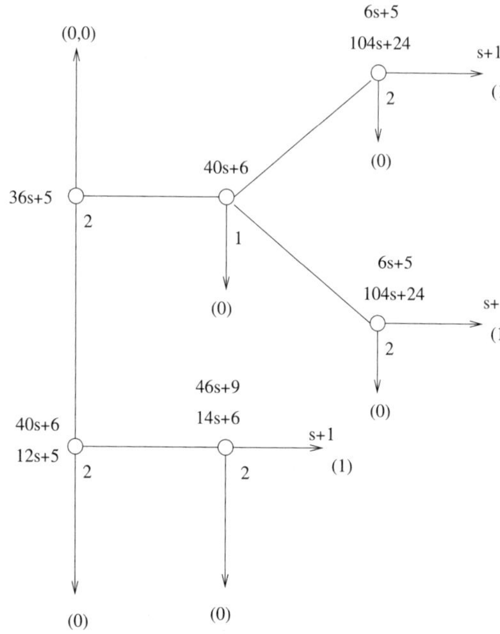

FIGURE 2.

The pull-back form $\bar{\omega}_{2,1}$ in its corresponding coordinates is given by $\bar{\omega}_{2,1} = \bar{y}_1^2\bar{y}_2^4 d\bar{y}_1 \wedge d\bar{y}_2 \wedge d\bar{z}_1$. Since $\bar{h}_{2,1}$ is given $\bar{y}_1^{20}\bar{y}_2^{12}(\bar{z}_1^2 - \bar{y}_1^3\bar{y}_2^2)$ and its non-degenerate then the set of candidates poles for the pair $(\bar{h}_{2,1}, \bar{\omega}_{2,1})$ is the union of $\{(20s+3), (12s+5)\}$ (from the previous polygon) and $\{(46s+9), (14s+6), (s+1)\}$. All these candidate poles are encoded in the bottom part of the diagram.

In the top part of the diagram we consider the pull-back form $\bar{\omega}_{1,1}$ given by $\bar{\omega}_{1,1} = y_1^4 dy_1 \wedge dy_2 \wedge dz_1$ (up to a unit). For $(\bar{h}_{1,1}, \bar{\omega}_{1,1})$ we get as candidate poles $\{(36s+5), (40s+6)\}$ (recall that we have only one compact dimensional face). This compact face have; since they behave in a similar way, we deal with one of them. The pull back $\tilde{\omega}$ of $\bar{\omega}_{1,1}$ under π is $\tilde{\omega} = w_1^5 dw_1 \wedge dw_2 \wedge d\tilde{z}_2$. Since $\tilde{h}$ is non-degenerate then set of candidate poles of $(\tilde{h}, \tilde{\omega})$ is $\{(40s+6), (6s+5), (104s+24), (s+1)\}$.

REMARK 3.30. Let $h \in k[[\mathbf{x}]][z]$ be a quasi-ordinary function with $h(\mathbf{0}, 0) = 0$ such that $D_z(h) = x_1^{\alpha_1} x_2^{\alpha_2} \dots x_d^{\alpha_d} \varepsilon(\mathbf{x})$, where $\varepsilon \in k[[\mathbf{x}]]$, $\varepsilon(\mathbf{0}) \neq 0$. After relabeling the variables x_i, one of the following conditions occurs. Either $\mathbf{ev}(h) = 0$, therefore $\alpha_i = 0$ for all $i = 1, \dots, d$ (which is is equivalent to $h = z^{\varepsilon}$, $\varepsilon = 0, 1$, in good coordinates) or $\alpha_k > 0$ for all $k = 1, \dots, m \leq d$ and $\alpha_k = 0$ for all $k \in \{m+1, \dots, d\}$. In such a case the last $d - m$ coordinates of any characteristic exponents of f always are zero. In particular compact faces of $\Gamma(h)$ are contained in the $(m+1)$-dimensional coordinate plane $x_{m+1} = \dots = x_d = 0$. The Newton maps never involve the last $d - m$ coordinates and therefore all compact faces of the new Newton polyhedra just only depend on the pull-back under the Newton maps of the coordinates $x_1, \dots, x_m, z$.

LEMMA 3.31. *The set of candidate poles does not depend on η, i.e., it always equals the result obtained for P-good coordinates.*

PROOF. This is essentially a consequence of the fact that the candidate poles can be read from the tree defined in 3.26; note that the multiplicities of the differential form can be deduced from the decorated tree.

Let us fix a choice η such that at a given stage we have chosen good coordinates $\mathbf{x}, z$ which are not P-good coordinates; we could consider another choice η' such that at this level, we have coordinates $\mathbf{x}, z_1$ which are P-good coordinates. This fact happens because

- z is not a factor of f;
- there is and edge γ in the Newton polygon for $\mathbf{x}, z_1$ such that
$$f_\gamma = w\mathbf{x}^{\mathbf{N}} \prod_{i=1}^{r} (z - a_i \mathbf{x}^{\mathbf{n}})^{n_i},$$
$w, a_i \in \mathbb{C}^*$, $n_i \in \mathbb{P}$, $\mathbf{N}, \mathbf{n} \in \mathbb{N}^d$, $r \geq 1$;
- this edge does not intersect the hyperplane $z = 0$.

The coordinate z_1 for η' is obtained by a generic change $z_1 = z + a\mathbf{x}^{\mathbf{n}}$; in the new Newton polygon, the corresponding edge γ' intersects the hyperplane $z = 0$. The sets of candidate poles obtained for the edges γ and above (in the choice η) and for the whole set of edges in the choice η' are the same.

For the choice η one must consider the candidate poles corresponding to the edges below γ. These candidate poles appear also in the choice η' but further in the Newton process, see [**4**]. The Newton mapping for these edges are composition of several Newton mappings in the choice η'. □

REMARK 3.32. In Remark 3.9 we have described how the method that we are using is affected for being in P-good coordinates or in good coordinates. The following diagrams show the two possible diagramas of $f(x,z) = (z^2 - x^5)(z - x^2)(z^3 - x^2)$ depending if we are in good coordinates (LHS) or P-good coordinates (RHS). One can say that one finds in both cases vertices with the same numerical data which will provide the same set of candidate poles.

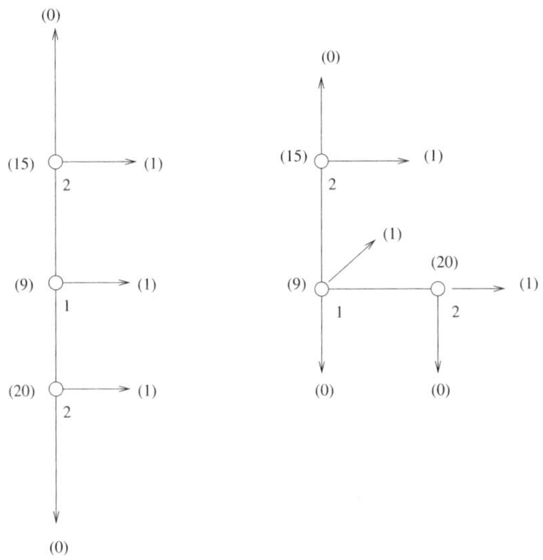

FIGURE 3.

DEFINITION 3.33. We will say that two quasi-ordinary power series with recursive choices η, η' of good coordinates *have the same* (η,η')*-characteristic exponents* if their graphs are equal. We will drop the term (η,η') if P-good coordinates is the common choice or if (η,η') are clear in the context.

Next lemma states the behavior of these graphs under the Newton mappings.

LEMMA 3.34. *Let $f \in k[[\mathbf{x}]][z]$ be a quasi-ordinary power series with a recursive choice η of good coordinates. Let $g \in k'[[\mathbf{x}]][z]$ be another quasi-ordinary power series (maybe defined over distinct algebraically closed fields of characteristic zero k') with a recursive choice η'. Then f and g have the same (η,η')-characteristic exponents if and only if*

(1) $\Lambda_{ND}(f) = \Lambda_{ND}(g)$,
(2) *there exists a bijection between the roots of the polynomials $f_{\boldsymbol{\lambda}_q}$ and $g_{\boldsymbol{\lambda}_q}$ and*
(3) *under this bijection if the Newton component f_j^q (or the root $f_{q,j}$) corresponds to the Newton component g_j^q (or the root $g_{q,j}$) then the quasi-ordinary power series $\bar{f}_{q,j}$ and $\bar{g}_{q,j}$ have the same (η,η')-characteristic exponents.*

It is possible to prove that the data η-characteristic exponents, that is the tree defined in 3.26, is the same as the characteristic exponents set of each irreducible component of f and the order of coincidence between distinct irreducible factors of f (as in the case of plane curves).

REMARK 3.35. Under any linear change of coordinates of type $z = \beta_{d+1}z_1$, $x_l = \beta_l y_l$, $l = 1\dots,d$, with $\beta_1,\dots,\beta_{d+1} \in \mathbf{G}_{m,k}$, the characteristic exponents, resp. monomials, of f do not change.

5. Transversal sections of a quasi-ordinary power series

Consider a quasi-ordinary $h \in k[[\mathbf{x}]][z]$ with $h(\mathbf{x},z) = x_1^{N_1}x_2^{N_2}\dots x_d^{N_d}g(\mathbf{x},z)$, where no x_i divides $g(\mathbf{x},z)$ and $N_l \geq 0$ for any $l = 1,\dots,d$. As usual we write $h(\mathbf{x},z) = x_1^{N_1}x_2^{N_2}\dots x_d^{N_d}f(\mathbf{x},z)u(\mathbf{x},z)$, $f(\mathbf{x},z)$ being the quasi-ordinary Weierstrass z-polynomial of degree n in $k[[\mathbf{x}]][z]$ whose roots are centred at $(\mathbf{0},0)$ and $u(\mathbf{0},0) \neq 0$. Assume that h is given in good coordinates. Suppose that a pair (h,ω) satisfying the support condition 1.3 is given, where $\omega = \prod_{j=1}^{d} x_j^{\nu_j-1}dx_1\wedge\dots\wedge dx_d\wedge dz$, $\nu_j \geq 1$.

Fix $i \in \{1,\dots,d\}$, and suppose that $d \geq 2$. Let k_i denote an algebraic closure of the fraction field of the formal power series ring $k[[x_i]]$. The i-th transversal section of h will be a finitely many disjoint union of k_i-quasi-ordinary formal power series obtained from h and their corresponding differential forms satisfying the support condition 1.3. Let $\mathbf{0}_i$ denote the d-tuple which has all coordinates but the i-th one, which is x_i, equal 0. Consider the polynomial $f(\mathbf{0}_i,z)$ over k_i of degree n :

$$f(\mathbf{0}_i,z) = z^n + \text{ lower degree terms } = \prod_{m=1}^{v_i}(z-\alpha_m)^{\delta_m} \in k_i[z],$$

$\alpha_m \in k_i$ and $n = \sum \delta_m$. By the Jung-Abhyankar Theorem for each root α_m of $f(\mathbf{0}_i,z) = 0$ in $k_i[z]$ there exists a k_i-quasi-ordinary formal power series $h_i^{\alpha_m} \in k_i[[\widehat{\mathbf{x}}_i]][z]$ centred at $(\widehat{\mathbf{0}}_i,\alpha_m)$, $\widehat{\mathbf{x}}_i$ being all variables but x_i.

DEFINITION 3.36. The *i-transversal section* of h consists in the set $\{h_i^{\alpha_m} : m = 1,\dots,v_i\}$ of k_i-quasi-ordinary formal power series, $h_i^{\alpha_m}$ centred at $(\widehat{\mathbf{0}}_i,\alpha_m)$. It is clear that $\mathbf{ev}(h_i^{\alpha_m}) < d$. If the root $\alpha_m \neq 0$ then we need a translation of type $z_1 = z - \alpha_m$ to describe the corresponding $h_i^{\alpha_m}$. On the other hand, if x_i is not an essential variable for h then the only root is $\alpha = 0$.

Since we are also interested in keeping the differential form, the *i-transversal section* of a pair (h,ω) consists in the set of finitely many pairs $(h_i^{\alpha_m},\omega_i)$ where $\omega_i = \prod_{j=1,j\neq i}^{d} x_j^{\nu_j-1}dx_1\wedge\dots\wedge\widehat{dx_i}\wedge\dots\wedge dx_d\wedge dz$.

EXAMPLE 3.37. Consider $f(x_1,x_2,z) = ((z^2-x_1^3)^2+x_1^7)^3 + x_1^{25}x_2^3$. For $i = 1$, the 1-transversal sections are the k_1-quasi-ordinary power series centred at the four roots of $((z^2-x_1^3)^2+x_1^7)^3 = 0$ in k_1. In fact these roots can be found using the Newton-Puiseux algorithm in k_1 for the previous polynomial. If α is one of these roots and $\tilde{z} := z - \alpha$ then the translated quasi-ordinary power series is

$$\tilde{z}^3u(x_1,\tilde{z}) + (x_1^{25})x_2^3, \quad u(0,0) \neq 0.$$

Its Newton polygon is the projection over the plane $x_1 = 0$ of the result of applying several Newton maps involving x_1, z to f.

For $i = 2$, the 2-transversal section is the k_2-quasi-ordinary power series $f_2^0 = ((z^2-x_1^3)^2+x_1^7)^3 + (x_2^3)x_1^{25} \in k_2[x_1,z]$ centred at the root $\alpha = 0$. In fact it is clear that the Newton polyhedron of f_2^0 is the projection over the plane $x_2 = 0$ of $\Gamma(f)$.

A description of the components h_i^α of the i-transversal section is as follows. Assume that

$$f(\mathbf{x},z)=\prod_{j=1}^{n}\left(z-\zeta_j(x_1^{1/m},\ldots,x_d^{1/m})\right), \tag{3.17}$$

with $\zeta_j(\mathbf{0})=0$. If $\zeta_j\neq 0$, since f is given in good coordinates, then

$$\zeta_j(\mathbf{x})=\alpha_1^{(j)}\mathbf{x}^{\boldsymbol{\lambda}_{(j)}}+\sum_{\boldsymbol{\lambda}_{(j)}<\boldsymbol{\lambda}}\alpha_{\boldsymbol{\lambda}}^{(j)}\mathbf{x}^{\boldsymbol{\lambda}},\ (\alpha_1^{(j)}\neq 0).$$

The characteristic exponents of type $(\mathbf{0},\frac{b_i}{n_1},\mathbf{0})\in\mathbb{Q}^d$, correspond to 1-dimensional faces in $CF(f)$ contained in the (x_i,z)-plane. We write in a unique way $\zeta_j(\mathbf{x})$ as sum of two formal power series $p_j(x_i^{1/m})+t_j(\mathbf{x})$ such that $p_j(x_i^{1/m})\in k[[x_i^{1/m}]]$ is the sum of monomials of ζ_j depending only on x_i and $t_j=\zeta_j-p_j(x_i)$. It is clear that $t_j(\mathbf{0}_i)=0$.

If $\alpha\in k_i$ is a root of $f(\mathbf{0}_i,z)=0$, $h_i^\alpha(\widehat{\mathbf{x}}_i,z)$ is

(3.18)

$$h_i^\alpha=u(\mathbf{x},z)\prod_{l=1}^{d}x_l^{N_l}\prod_{j:p_j(x_i^{1/m})=\alpha}\left(z-(\alpha+t_j(x_1^{1/m},\ldots,\widehat{x}_i,\ldots,x_d^{1/m})\right)\in k_i[[\widehat{\mathbf{x}}_i]][z],$$

where t_j is seen as an element in $k_i[[x_1^{1/m},\ldots,\widehat{x}_i,\ldots,x_d^{1/m}]]$. The last product gives a decomposition of h_i^α into its irreducible roots too, because each factor has z-degree one. To work with h_i^α we first perform, if necessary, the translation $z_1=z-\alpha$, $\alpha\in k_i$. Such a translation is a composition of some Newton maps of h associated with their corresponding Newton components (all of them of type $(\mathbf{0},a_i/w_i,\mathbf{0})$) and some change of variables in order to obtain good coordinates. The Newton map π_1 associated with a Newton component of h of type $(\mathbf{0},b_i/n_1,\mathbf{0})$ is given by $x_i=y_i^{p_i}$, $z=(z_1+\alpha)y_i^{\bar{b}_i}$ and $x_j=y_j$, if $j\neq i$. After π_1, we need a translation of type $z_1=\tilde{z}-\phi(\mathbf{x})$ to get good coordinates. This translation does not change the position of the i-th variable. If in order to find α we will need eventually, more Newton maps of type $(\mathbf{0},a_i/w_i,\mathbf{0})$. We finish when no more characteristic exponent of the mentioned type appear in the process to get α. Therefore the composition of these Newton maps and their corresponding translations, which we denote by π, is the same as the Newton-Puiseux algorithm to find the root α of $f(\mathbf{0}_i,z)$ in $k_i[z]$. The map π does nothing onto the set of variables different from i-th one. Let $\tilde{f}$ (resp. $\tilde{h}$) denote the pull-back of f (resp. h) under the map π. Up to a factor $x_i^{k/m}$, $\tilde{f}(\mathbf{x},z)=\prod_{j=1}^{n}\left(z-t_j(x_1^{1/m},\ldots,x_d^{1/m}\right)$ and

$$h_i^\alpha(\widehat{\mathbf{x}}_i,z)=\tilde{u}\prod_{l=1}^{d}x_l^{N_l}u(\mathbf{x},z)\prod_{j:p_j(x_i^{1/m})=\alpha}\left(z-t_j(x_1^{1/m},\ldots,\widehat{x}_i,\ldots,x_d^{1/m})\right).$$

Moreover, ω_i and the pull-back $\pi^*\omega$ are the same, up to a constant, as differential forms over k_i.

Once this translation is done, the Newton polyhedron $\Gamma_i(h_i^\alpha)$ of h_i^α as k_i-quasi-ordinary power series is the projection over the hyperplane $x_i=0$ of the Newton polyhedron of the pull-back $\tilde{f}$; this is clear for $\alpha=0$ and for the other case is similar. In particular $\Gamma_i(h_i^\alpha)$ is a monotone polygonal path. In fact, the k_i-quasi-ordinary power series h_i^α is in good coordinates because the condition not to be a power of a linear form is generic.

REMARK 3.38. Note that if h is given in P-good coordinates, we cannot deduce that h_i^α is also in P-good coordinates. This is the main reason for introducing the more general concept of good coordinates.

EXAMPLE 3.39. Let us consider several examples concerning good coordinates, Newton polygons and transversal sections.

(1) Consider $f(x_1, x_2, z) = (z^2 - x_1^2 x_2^5)(z^2 - x_1^3 x_2^7)$ which is given in P-good coordinates. The 2-transversal section is not in P-good coordinates.
(2) Consider $f(x_1, x_2, z) = (z^2 - x_1^2 x_2^5)(z^2 - x_1^2 x_2^7)$. The Newton polygon $\Gamma(f)$ has two edges which are projected over only one edge if we project over $x_2 = 0$. The 2-transversal section has at least two different components with the same characteristic exponent.
(3) Consider$f(x_1, x_2, z) := (z^2 - x_1^2 x_2^5)^2 + x_1^4 x_2^{17}$. The 2-transversal section is $f_2^0 = z^4 - (2x_2^5)x_1^2 z^2 + (x_2^{10} + x_2^{17})x_1^4$. In this case some more monomials of x_1^4 have appeared. These monomials are hidden in $ND(f)$.

To sum up:

PROPOSITION 3.40. *With the above notations,*

(1) *The formal power series* $h_i^{\alpha_m} \in k_i[[\widehat{\mathbf{x}}_i]][z]$ *centred at* $(\widehat{\mathbf{0}}_i, \alpha_m)$, *defines a quasi-ordinary power series with* $\mathbf{ev}(h_i^{\alpha_m}) < d$.
(2) *The Newton process for* h *induces Newton process for* $h_i^{\alpha_m}$. *Each step of the Newton process for* h *is necessary for at least one* $h_i^{\alpha_m}$, $i \in \{1, \ldots, d\}$, *and* $m \in \{1, \ldots, v_i\}$.
(3) *Fix* $i \in \{1, \ldots, d\}$ *and* $m \in \{1, \ldots, v_i\}$. *Then*
   i) *If* $\alpha_m = 0$ *then the quasi-ordinary power series* h_i^0 *is in good coordinates.*
   ii) *At each step of the Newton process for* h, $\Gamma(h_i^0)$ *is the projection of* $\Gamma(h)$ *on the hyperplane* $x_i = 0$.
   iii) *If* $\alpha_m \neq 0$, *then some steps of the Newton process for* h *firstly provide the translation* $z = z_1 - \alpha_m$ *we need to put* α_m *at zero. After that we apply i) and ii).*
   iv) *Every pair* $(h_i^{\alpha_m}, \omega_i)$ *satisfies the support condition 1.3.*
(4) *In fact*

$$CP(h, \omega) \subset \bigcup_{i=1}^{d} \bigcup_{m=1}^{v} CP(h_i^{\alpha_m}, \omega_i), \tag{3.19}$$

Recall that the second union runs over all roots of $f(\mathbf{0}_i, z) = 0$ *in* $k_i[z]$.

The only fact in the proposition that is not yet proved is the inclusion (3.19). The set $CP(h, \omega)$ of *candidate poles* for (h, ω) can be written as $CP(h, \omega) = CP_1(h, \omega) \cup \bigcup CP(\bar{h}_{q,j}, \bar{\omega}_{q,j})$, see definition 3.28, where the set $CP_1(h, \omega)$ is completely described from $ND(h)$ and ω. Since this monotone polygonal path is determined by (and determines) its projections over all hyperplanes $\{x_i = 0\}, i = 1, \ldots, d$, then $CP_1(h, \omega) = \bigcup_{i=1}^d CP_1(h_i^0, w_i)$. In the previous identity we forget the roots which are not centred at zero because some Newton maps are needed to find them.

In general, (3.19) follows from the fact that candidate poles are determined by the characteristic monomials of the quasi-ordinary power series. The characteristic monomials are determined by differences of its roots. In particular for each of the i-transversal section the difference of two of its roots in $k_i[[x_1^{1/m}, \ldots, \widehat{x_i}, \ldots, x_d^{1/m}]]$ can be computed as the specialization of the difference of the corresponding roots in $k[[x_1^{1/m}, \ldots, x_i{}^{1/m}, \ldots, x_d^{1/m}]]$.

In fact we can say more. Fix a characteristic exponent of f and a Newton component with such a characteristic exponent. Fix $i \in \{1, \ldots, d\}$. If the characteristic exponent is $(\mathbf{0}, b_i/n_1, \mathbf{0})$ then we have already described what is going on with the Newton map π associated with the Newton component. Otherwise, reading the characteristic exponents and the roots from equations (3.17) and (3.18), the characteristic exponents of the i-transversal section of $f \circ \pi$ are the same as those of the pull-back under the corresponding Newton map for the i-transversal section. In particular at each step the Newton polyhedron of the i-transversal section is the projection of the Newton polyhedron of h.

The fact that we get the same differential form, up to a constant factor, in x_i is clear because this is a general fact not related to quasi-ordinary formal power series.

CHAPTER 4

Denef-Loeser motivic zeta function under the Newton maps

We keep the field k being algebraically closed and of characteristic zero. We denote an algebraic closure of the quotient field $k((t))$ of the domain $R := k[[t]]$ by K. The goal is to compute zeta functions for a k-quasi-ordinary power series using induction on its depth. In Section 3.3 we have computed the A-part. In order to compute the B-part we must take into account that a quasi-ordinary power series may be degenerate with respect to its Newton polyhedron. The main idea is to use Newton maps to measure the remaining arcs in terms of the quasi-ordinary power series obtained after these Newton maps. The problem is the existence of some arcs which cannot be lifted under such maps. We need to combine Newton maps and $k[t]$-morphisms to solve this problem. Therefore we introduce technical objects, called $\mathcal{W}$ and $\widetilde{\mathcal{W}}$-quasi-ordinary series in order to have families of power series closed under Newton maps to deal with. We will compute the Denef-Loeser motivic zeta functions for these series following the ideas of Sections 3.2 and 3.3; the computation of the A-part is similar to the one in Section 3.3 but there are some small differences and we will compute inductively the whole zeta function.

DEFINITION 4.1. Let $h^t(\mathbf{x}, z) \in R[[x_1, \ldots, x_d]][z]$ be a formal power series such that $h^t = t^\theta x_1^{N_1} x_2^{N_2} \ldots x_d^{N_d} f^t(\mathbf{x}, z) u^t(\mathbf{x}, z)$ where x_j does not divide $f^t(\mathbf{x}, z)$ $\forall j = 1, \ldots, d$, t does not divide $f^t(\mathbf{x}, z)u^t(\mathbf{x}, z)$, $\theta \in \mathbb{N}$ and $u^t(\mathbf{0}, 0) \neq 0$. We say that h^t is *$\mathcal{W}$-quasi-ordinary* if its z-discriminant is

$$D_z(f^t) = t^\beta x_1^{\alpha_1} x_2^{\alpha_2} \ldots x_d^{\alpha_d} \varepsilon(t, \mathbf{x}), \tag{4.1}$$

where $\varepsilon(0, \mathbf{0}) \neq 0$; therefore if we consider t as a variable, $f^t \in k[[t, \mathbf{x}]][z]$ is a k-quasi-ordinary power series. Let $J \subset \{1 \ldots, d\}$ and let J' be its complement; we say that h^t is *J-bounded* if its roots ζ^t satisfy

$$\zeta^t \in k[\mathbf{x}_J^{1/m}][[\mathbf{x}_{J'}^{1/m}, t]], \text{ for some } m. \tag{4.2}$$

REMARK 4.2. The main property of a J-bounded $\mathcal{W}$-quasi-ordinary series h^t is that $\forall \mathbf{y}_J^0 \in \mathbb{C}^d$ such that the entries in J' are zero the series $h^t(\mathbf{x} + \mathbf{y}_J^0, z)$ is a well-defined $\mathcal{W}$-quasi-ordinary series.

Let $f(\bar{\mathbf{x}}, \bar{z}) \in k[\bar{x}_1, \ldots, \bar{x}_d][\bar{z}]$ be a quasi-ordinary polynomial whose roots only contain the (finite set of) characteristic monomials of f^t as K-quasi-ordinary power series in a new set of variables $\bar{x}_1, \ldots, \bar{x}_d, \bar{z}$. Let $h := \bar{x}_1^{N_1} \ldots \bar{x}_d^{N_d} f(\bar{\mathbf{x}}, \bar{z})$. In particular the K-quasi-ordinary power series h^t and the k-quasi-ordinary power series h have the same characteristic exponents and the same characteristic monomials.

If $f^t(\mathbf{x}, z) \in k[[t]][[\mathbf{x}]][z]$ is not in *good* coordinates, by Remark 3.8, the change of coordinates to put $f^t(\mathbf{x}, z)$ in good coordinates is an automorphism of $k[[t]][[\mathbf{x}]][z]$ of type $z \mapsto z + m^t(\mathbf{x})$, $m \in k[[t]][[\mathbf{x}]]$. In particular the t-variable does not change and condition (4.1) on the discriminant is preserved. It is easily seen that if h^t is J-bounded then condition (4.2) is also preserved because $m^t(\mathbf{x})$ comes from some monomials of the roots of h^t. Then we assume f^t and $f(\bar{\mathbf{x}}, \bar{z})$ are in good coordinates.

Consider differential forms $\omega = \prod_{j=1}^d x_j^{\nu_j - 1} d\mathbf{x} \wedge dz$ and $\bar{\omega} = \prod_{j=1}^d \bar{x}_j^{\nu_j - 1} d\bar{\mathbf{x}} \wedge d\bar{z}$, with $\nu_j \geq 1$. From now on we assume that $(h, \bar{\omega})$ satisfies the support condition 1.3. Let $X := \mathbb{A}_k^{d+1}$ and set

$$V_{n,m}^t := \{\varphi \in \mathcal{L}_0(X) : \operatorname{ord}(h^t \circ \varphi) = n + \theta,\ \operatorname{ord}(\prod_{j=1}^d x_j^{\nu_j - 1} \circ \varphi) = m\}; \tag{4.3}$$

define $V_{n,m}$ in the same way for h (forgetting θ). The sets $V_{n,m}^t$ are measurable because of the support condition. We define the power series

$$\begin{aligned} Z_{DL}^{\mathcal{W}}(h^t, \omega, T) &:= T^\theta \sum_{n \in \mathbb{N}} \left(\sum_{m \in \mathbb{N}} \mathbb{L}^{-m} \mu_X(V_{n,m}^t) \right) T^n \in \widehat{\mathcal{M}}_k[[T]], \\ Z_{\mathrm{top},0}^{\mathcal{W}}(h^t, \omega, s) &:= \chi_{\mathrm{top}}(Z_{DL}^{\mathcal{W}}(h^t, \omega, \mathbb{L}^{-s})). \end{aligned} \tag{4.4}$$

Recall that

$$\begin{aligned} Z_{DL}(h, \bar{\omega}, T) &= \sum_{n \in \mathbb{N}} \left(\sum_{m \in \mathbb{N}} \mathbb{L}^{-m} \mu_X(V_{n,m}) \right) T^n, \\ Z_{\mathrm{top},0}(h, \bar{\omega}, s) &= \chi_{\mathrm{top}}(Z_{DL}(h, \bar{\omega}, \mathbb{L}^{-s})). \end{aligned}$$

We know from the results of Denef and Loeser that $Z_{DL}(h, \bar{\omega}, T)$ is a rational function. The last part of the section is devoted to proving the rationality of $Z_{DL}^{\mathcal{W}}(h^t, \omega, T)$ and to provide a *small set* of candidate poles for such functions, see Definition 3.28. Recent interesting results of J. Sebag [**41**] prove the rationality of this power series.

THEOREM 4.3. *If h^t is $\mathcal{W}$-quasi-ordinary then the zeta function*

$$Z_{DL}^{\mathcal{W}}(h^t, \omega, T) \in \mathbb{Z}\left[\mathbb{L}, \mathbb{L}^{-1}, \left(1 - \mathbb{L}^{-\nu} T^N\right)^{-1}\right][T]$$

where $(N, \nu) \in CP(h, \bar{\omega})$. Moreover, $Z_{\mathrm{top},0}^{\mathcal{W}}(h^t, \omega, s)$ exists.

We need to define a subfamily of $\mathcal{W}$-quasi-ordinary series in order to obtain more information on the local topological zeta function.

DEFINITION 4.4. A symbol $\widetilde{\mathcal{W}} := (p; g_1, \ldots, g_d)$, where $p \in \mathbb{P}, g_i \in \mathbb{N}$, $i = 1, \ldots, d$, is called a *weight*. The semigroup $\Gamma_{\widetilde{\mathcal{W}}} \subset \mathbb{N}^d$ associated with $\widetilde{\mathcal{W}}$ is the additive set of $\boldsymbol{\alpha} \in \mathbb{N}^d$ such that

$$w_{\widetilde{\mathcal{W}}}(\boldsymbol{\alpha}) := \sum_{i=1}^d \frac{\alpha_i g_i}{p} \in \mathbb{N}.$$

DEFINITION 4.5. Let $h^t(\mathbf{x}, z) \in R[[x_1, \ldots, x_d]][z]$ be a formal power series such that $h^t = t^\theta x_1^{N_1} x_2^{N_2} \ldots x_d^{N_d} f^t(\mathbf{x}, z) u^t(\mathbf{x}, z)$ with x_j does not divide $f^t(\mathbf{x}, z)$ $\forall j = 1, \ldots, d$, $u^t(\mathbf{0}, 0) \neq 0$, t does not divide $f^t(\mathbf{x}, z) u^t(\mathbf{x}, z)$ and $\theta \in \mathbb{N}$. We call it $\widetilde{\mathcal{W}}$-quasi-ordinary if it is $\mathcal{W}$-quasi-ordinary and all monomials in h^t are of type $t^{w_{\widetilde{\mathcal{W}}}(\boldsymbol{\alpha})} \mathbf{x}^{\boldsymbol{\alpha}} z^n$, where $\boldsymbol{\alpha} \in \Gamma_{\widetilde{\mathcal{W}}}$ (and some non-zero coefficient in k).

REMARK 4.6. If h^t is $\widetilde{\mathcal{W}}$-quasi-ordinary then the power series $h^1 := h^t_{|t=1}$ is k-quasi-ordinary and has the same characteristic exponents as h^t. If h^t is $\widetilde{\mathcal{W}}$-quasi-ordinary, it is true also for f^t because the d-tuple $(N_1, \dots, N_d) \in \Gamma_{\widetilde{\mathcal{W}}}$ and $\theta = w_{\widetilde{\mathcal{W}}}(N_1, \dots, N_d)$. Moreover, if we put h^t in good coordinates, the new series is also $\widetilde{\mathcal{W}}$-quasi-ordinary.

REMARK 4.7. Let h^t be a $\widetilde{\mathcal{W}}$-quasi-ordinary power series, $\widetilde{\mathcal{W}} := (p; g_1, \dots, g_d)$. Then, if $J := \{j \mid 1 \leq j \leq d,\ g_j \neq 0\}$ then h^t is J-bounded.

DEFINITION 4.8. Let h^t be a $\widetilde{\mathcal{W}}$-quasi-ordinary power series. We say that h^t is *Newton compatible* if the number of irreducible factors (different from t) of $h^t_{|ND(h^t)}$ in R equals the number of irreducible factors of $h^1_{|ND(h^1)}$ in k. For instance $z^2 - tx^2y^2$ is not Newton compatible and it is $\widetilde{\mathcal{W}}$-quasi-ordinary for some adequate weight.

THEOREM 4.9. *If h^t is $\widetilde{\mathcal{W}}$-quasi-ordinary and Newton compatible then*

$$Z^{\mathcal{W}}_{\text{top},0}(h^t, \omega, s) = Z_{\text{top},0}(h, \bar{\omega}, s).$$

From now on we suppose that h^t is $\mathcal{W}$-quasi-ordinary; we will indicate explicitly when we consider it as $\widetilde{\mathcal{W}}$-quasi-ordinary. The rest of this section is devoted to the proof of Theorems 4.3 and 4.9. We will focus on the proof of Theorem 4.3 and we will point out the special arguments required for the proof of Theorem 4.9. Since the results do not depend on the factor T^θ, we do not take care of it. We also assume from now on that h^t (or f^t) is given in good coordinates. The proofs are given by induction on the depth, see Definition 3.24. Example 4.36 illustrates the main steps of the induction.

Step 1. depth$(h) = $ depth$(h^t) = 0$.

Then $ND(h)$ has only one compact face which is 0-dimensional,

$$Z^{\mathcal{W}}_{DL}(h^t, \omega, T) = \mathbb{L}^{-(d+1)}(\mathbb{L}-1)^{d+1} \frac{\mathbb{L}^{-(1+\sum \nu_j)} T^{1+\sum N_j}}{1 - \mathbb{L}^{-1}T} \prod_{j=1}^{d} \frac{1}{1 - \mathbb{L}^{-\nu_j} T^{N_j}},$$

and $Z^{\mathcal{W}}_{DL}(h^t, \omega, T) = Z_{DL}(h, \bar{\omega}, T)$. Then $Z^{\mathcal{W}}_{\text{top},0}(h^t, \omega, s) = Z_{\text{top},0}(h, \bar{\omega}, s)$.

Step 2. Assume that depth$(h) > 0$.

We keep the notation of Chapter 2. Let S be defined by $x_1 \cdots x_d z = 0$, (or $\bar{x}_1 \cdots \bar{x}_d \bar{z} = 0$); we consider arcs in $\varphi \in \mathcal{L}_0(X) \setminus \mathcal{L}_0(S)$ and define $\mathbf{k}(\varphi)$ and $\mathbf{a}(\varphi)$ as in Chapter 2.

The following easy remark will be a key point for understanding the dual decomposition associated with $f^t(\mathbf{x}, z) \in R[[\mathbf{x}]][z]$. Each $\varphi \in \mathcal{L}_0(X)$ uniquely determines an arc $\psi := (t, \varphi) \in \mathcal{L}_0(\mathbb{A}^1_k \times X)$ so that $\operatorname{ord}_t(f^t \circ \varphi) = \operatorname{ord}_t(g \circ \psi)$, where $g(y, \mathbf{x}, z) \in k[[y, \mathbf{x}]][z]$ is the same function as f^t but t is substituted by a new variable y.

Step 3. Newton polyhedron of $h \in k[\bar{\mathbf{x}}, \bar{z}]$.

This Newton polyhedron $\Gamma(h)$ and its dual decomposition have been described in Section 3.3. Denote by $\bar{\gamma}_1, \dots, \bar{\gamma}_r$ its 1-dimensional faces with corresponding vertices $\bar{\tau}_0, \bar{\tau}_1, \ldots, \bar{\tau}_r$. We assume that $\bar{\tau}_r$ is the $\bar{z}$-highest vertex as in Remark 3.13. Thus we can write

$$f_{ND(f)} = z^{\varepsilon} \prod_{q=1}^{r} \prod_{\bar{j}=1}^{v(q)} (f_{q,\bar{j}})^{m_{q,\bar{j}}} \text{ where } f_{q,\bar{j}} := \bar{z}^{n_1^q} - \bar{\beta}_{\bar{j}}^q \bar{x}_1^{b_1^q} \ldots \bar{x}_d^{b_d^q},$$

with $\varepsilon = 0$ (resp. 1) if $I = I'$ (resp. $I \neq I'$). Recall that in the dual space, the reduced integer equation of the hyperplane l^q is $\eta_q(v_1, \ldots, v_d, v_{d+1}) = 0$, see equation (3.5). The intersection of these hyperplanes with $\mathbb{R}_{>0}^{d+1}$ are the d-dimensional cones $\Delta_{\bar{\gamma}_q}$. They determine the $(d+1)$-dimensional cones $\Delta_{\bar{\tau}_q}$, $q = 0, 1, \ldots, r$. The cones associated with compact faces of $\Gamma(h)$ (or $\Gamma(f)$) give a partition of $\mathbb{R}_{>0}^{d+1}$ in the disjoint union $\bigcup \Delta_{\bar{\tau}}$.

If $\varphi \in \mathcal{L}_0(X) \setminus \mathcal{L}_0(S)$ and $\mathbf{k} := \mathbf{k}(\varphi)$, the order of the differential form $\bar{\omega}$ is $m(\mathbf{k}) := \sigma_{\bar{\omega}}(\mathbf{k}) - \mathbf{k}$. If $\mathbf{k} \in \mathbb{P}^{d+1}$ is fixed, the semialgebraic subset $V_{n,m}^{\mathbf{k}}$ is naturally defined. It is empty unless $m = m(\mathbf{k})$. Then for each compact face $\bar{\tau}$ define

$$Z_{DL}^{\bar{\tau}}(h, \bar{\omega}, T) := \sum_{\mathbf{k} \in \Delta_{\bar{\tau}} \cap \mathbb{P}^{d+1}} Z_{DL}^{\mathbf{k}}(T).$$

Step 4. Description of the Newton polyhedron $\Gamma^t(f^t)$ of $f^t \in k[[t, \mathbf{x}]][z]$.

The discriminant condition (4.1) implies that f^t is a k-quasi-ordinary power series in good coordinates in $(d+2)$ variables. Then its Newton polyhedron $\Gamma^t(f^t)$ has only 0 and 1-dimensional compact faces, say m edges $\tilde{\gamma}_1, \ldots, \tilde{\gamma}_m$ and the $m+1$ corresponding vertices. We can write

$$f_{ND(f^t)}^t = z^{\varepsilon} \prod_{p=1}^{m} \prod_{s=1}^{v(p)} (z^{\tilde{n}_1^p} - \beta_s^p t^{\tilde{b}_0^p} x_1^{\tilde{b}_1^p} \ldots x_d^{\tilde{b}_d^p})^{m_{p,s}}, \tag{4.5}$$

where $\gcd(\tilde{n}_1^p, \tilde{b}_0^p, \tilde{b}_1^p, \ldots, \tilde{b}_d^p) = 1$ and $\varepsilon = 0$ (resp. 1) if $I = I'$ (resp. $I \neq I'$).

The characteristic exponents associated with $\Gamma^t(f^t)$ are of type $(\frac{\tilde{b}_0^p}{\tilde{n}_1^p}, \boldsymbol{\lambda}_p^t) \in \mathbb{Q}^{d+1}$ where $\tilde{n}_1^p \boldsymbol{\lambda}_p^t = (\tilde{b}_1^p, \ldots, \tilde{b}_d^p) \in \mathbb{N}^d$. We need also to define $n_1^p, b_1^p, \ldots, b_d^p$ such that $\gcd(n_1^p, b_1^p, \ldots, b_d^p) = 1$ and $n_1^p \boldsymbol{\lambda}_p^t = (b_1^p, \ldots, b_d^p)$.

Given $\boldsymbol{\lambda}_q$ from a compact face $\bar{\gamma}_q$ of $\Gamma(h)$, $q = 1, \ldots, r$, we collect the set S_q of rationals $\frac{b}{n}$ such that there exists a compact face $\tilde{\gamma}_p$ of $\Gamma^t(h^t)$ whose associated characteristic exponent is $(\frac{b}{n}, \boldsymbol{\lambda}_q)$; all characteristic exponents of h^t are obtained in this way.

Let $u(q)$ be the cardinality of S_q; we denote by $\frac{\alpha_{q,j}}{n_1^{q,j}}$, $j = 1, \ldots, u(q)$, the elements of S_q, where $\alpha_{q,j} \in \mathbb{N}$ and $n_1^{q,j} = \tilde{n}_1^p$ for the corresponding $p \in \{1, \ldots, m\}$. We denote also $s_1^{q,j} := \frac{n_1^{q,j}}{n_1^q} \in \mathbb{N}$. Then identity (4.5) can be rewritten as follows

$$f_{ND(f^t)}^t = z^{\varepsilon} \prod_{q=1}^{r} \prod_{j=1}^{u(q)} \prod_{w=1}^{w(q,j)} ((z^{n_1^q})^{s_1^{q,j}} - \beta_w t^{\alpha_{q,j}} (x_1^{b_1^q} \ldots x_d^{b_d^q})^{s_1^{q,j}})^{m_{q,j,w}},$$

where $v(q) = \sum_{j=1}^{u(q)} s_1^{q,j} w(q,j)$ since f^t and f have by hypothesis the same characteristic exponents.

REMARK 4.10. If h^t is $\widetilde{\mathcal{W}}$-quasi-ordinary and Newton compatible then $r = m$, it means that for each $q = 1, \ldots, r$ we have $\#S_q = 1$ and $S_q \in \mathbb{N}$. In this case $s_1^{q,j} = 1$, $\tilde{b}_j^q = b_j^q$.

In the dual space $(\mathbb{R}^{d+2})^*$, with coordinates $(v_0, v_1, \ldots, v_d, v_{d+1})$, the dual decomposition induced by $\Gamma^t(h^t)$ is given by dual cones $\Delta_{\tilde{\gamma}_p}$ associated with the hyperplanes, $\tilde{l}^p : \tilde{b}_0^p v_0 + \tilde{b}_1^p v_1 + \ldots + \tilde{b}_d^p v_d - \tilde{n}_1^p v_{d+1} = 0$, for $p \in \{1, \ldots, m\}$, and dual cones $\Delta_{\tilde{\tau}}$ associated with a vertex $\tilde{\tau}$. We identify the affine hyperplane $H = \{v_0 = 1\}$ with $(\mathbb{R}^{d+1})^*$ and $H \cap (\mathbb{R}_+^{d+2})^*$ with $(\mathbb{R}_+^{d+1})^*$ and consider the decomposition induced by $\Delta_{\tilde{\gamma}_p}$ and $\Delta_{\tilde{\tau}}$ in $(\mathbb{R}_+^{d+1})^*$. This decomposition has the same properties as the initial one. It has exactly m ordered d-dimensional planes in H and $m+1$ *pieces* of dimension $(d+1)$ between them. Let us describe them better.

For $q = 1, \ldots, r$, and $j \in \{1, \ldots, u(q)\}$ define the cone $\Delta_{q,j}^t$ as the intersection with $\mathbb{R}_{>0}^{d+1}$ of the hyperplane $l_t^{q,j}$ with integral equation:

$$s_1^{q,j}\eta_q(v_1, \ldots, v_d, v_{d+1}) + \alpha_{q,j} = 0, \text{ see (3.5)}. \tag{4.6}$$

Observe that $m = \sum_{q=1}^r u(q)$.

The $(d+1)$-dimensional convex rational polyhedra Δ_s^t, $s = 1, \ldots, m-1$, are either the region $M_t^{q,j}$ (that we will call of type M) contained between two parallel hyperplanes $l_t^{q,j}$ and $l_t^{q,j+1}$ or the region N_t^q (of type N) contained between the hyperplanes $l_t^{q,u(q)}$ and $l_t^{q+1,1}$ for some $q = 1, \ldots, m-1$. The first one Δ_0^t and the last one Δ_m^t are

$$\begin{aligned} \Delta_0^t &:= \{v \in \mathbb{R}_{>0}^{d+1} : s_1^{1,1}\eta_1(v_1, \ldots, v_d, v_{d+1}) + \alpha_{1,1} < 0\}, \\ \Delta_m^t &:= \{v \in \mathbb{R}_{>0}^{d+1} : s_1^{r,u(r)}\eta_r(v_1, \ldots, v_d, v_{d+1}) + \alpha_{r,u(r)} > 0\}. \end{aligned} \tag{4.7}$$

Thus we fix the partition of $\mathbb{R}_{>0}^{d+1}$ (identified with H) as disjoint union of three distinct types of convex rational polyhedra Δ_τ^t:

- d-dimensional cones $\Delta_{q,j}^t$ corresponding to the hyperplanes $l_t^{q,j}$,
- convex rational polyhedra of type N, and
- convex rational polyhedra of type M.

Following the conventions of this work the convex rational polyhedra of type M or N will be called *vertices* and those of the first type will be called an *edge*.

Let $\varphi \in \mathcal{L}_0(X) \setminus \mathcal{L}_0(S)$ and let $\mathbf{k} := \mathbf{k}(\varphi)$. The order of the differential form ω is the same as for $\bar{\omega}$, i.e., $m(\mathbf{k}) = \sigma_\omega(\mathbf{k}) - \mathbf{k}$. Fix $\mathbf{k} \in \mathbb{P}^{d+1}$ and define

$$V_{n,m}^{\mathbf{k},t} := \{\varphi \in \mathcal{L}_0(X) \backslash \mathcal{L}_0(S) : \mathbf{k}(\varphi) = \mathbf{k}, \operatorname{ord}(h^t \circ \varphi) = n + \theta, \operatorname{ord}(\prod_{j=1}^d \bar{x}_j^{\nu_j - 1} \circ \varphi) = m\};$$

These sets are empty unless $m = m(\mathbf{k})$, then we define

$$Z_{DL}^{\mathbf{k},\mathcal{W}}(T) := \sum_{n \in \mathbb{N}} \mathbb{L}^{-m(\mathbf{k})} \mu_X(V_{n,m(\mathbf{k})}^{\mathbf{k},t}) T^n.$$

For each convex rational polyhedron Δ_τ^t in the previous partition, define

$$Z_{DL}^{\Delta_\tau^t,\mathcal{W}}(h^t, \omega, T) := \sum_{\mathbf{k} \in \Delta_\tau^t \cap \mathbb{P}^{d+1}} Z_{DL}^{\mathbf{k},\mathcal{W}}(T).$$

REMARK 4.11. Each hyperplane l_q is parallel to the hyperplanes $l_t^{q,j}$. The region associated with a vertex $\bar{\tau}_q$ is a cone delimited by two hyperplanes which are parallel to the hyperplanes delimiting N_t^q (different from the coordinate hyperplanes).

REMARK 4.12. If h^t is $\widetilde{\mathcal{W}}$-quasi-ordinary and Newton compatible then no region of type M exists.

Step 5. Newton polyhedron of h^t as a K-quasi-ordinary function.

This Newton polyhedron $\Gamma^K(h^t)$ of h^t is the projection over the hyperplane $t = 0$ of the Newton polyhedron of h^t and it coincides with $\Gamma(h)$. For $q \in \{1, \ldots, r\}$, we define

$$\widehat{f}^t_{\boldsymbol{\lambda}_q} := \prod_{j=1, \alpha_{q,j} \in S_q}^{u(q)} \widehat{f}^t_{q,j}, \text{ where } \widehat{f}^t_{q,j} := \prod_{w=1}^{w(q,j)} ((z^{n_1^q})^{s_1^{q,j}} - \beta_w t^{\alpha_{q,j}} (x_1^{b_1^q} \ldots x_d^{b_d^q})^{s_1^{q,j}})^{m_{q,j,w}}.$$

1. Vertices of the dual decomposition

In the following, the aim is to compute $Z_{DL}^{\Delta^t_\tau, \mathcal{W}}(h^t, \omega, T)$ for τ a vertex of type M or N and compare it with the computations we have already done for $Z^{\bar{\tau}}_{DL}(h, \bar{\omega}, T)$, see (3.15).

Take a convex rational polyhedron Δ^t_τ of the dual decomposition of $\mathbb{R}^{d+1}_+$ which belongs to any of the two types M or N. It corresponds to a monomial $t^\delta \mathbf{x}^{\boldsymbol{\alpha}} z^n$ whose support is the intersection of two 1-dimensional compact faces of $\Gamma^K(h^t)$. The cone Δ^t_τ is the positive region delimited by two inequalities, say

$$\tilde{\eta}_1(v_1, \ldots, v_d, v_{d+1}) + \tilde{\alpha}_1 > 0,$$
$$\tilde{\eta}_2(v_1, \ldots, v_d, v_{d+1}) + \tilde{\alpha}_2 < 0,$$

where $\tilde{\eta}_1$ and $\tilde{\eta}_2$ can define parallel (or not) hyperplanes. Let $\varphi \in \mathcal{L}_0(X) \setminus \mathcal{L}_0(S)$ and let $\mathbf{k} := \mathbf{k}(\varphi)$ and $n_\tau(\mathbf{k}) := \alpha_1 k_1 + \cdots + \alpha_d k_d + n k_{d+1}$. We have $\operatorname{ord}(h^t \circ \varphi) = n_\tau(\mathbf{k}) + N(\mathbf{k}) + \delta$, with $N(\mathbf{k}) := \sum_{j=1}^d N_j k_j$ (we are forgetting the exponent θ).

LEMMA 4.13. *If* $\mathbf{k} \in \Delta^t_\tau$, *then*

$$\mathbb{L}^{-m(\mathbf{k})} \mu_X(V^{\mathbf{k},t}_{n_\tau(\mathbf{k})+\delta+N(\mathbf{k}), m(\mathbf{k})}) = \mathbb{L}^{-(d+1+\sigma_\omega(\mathbf{k}))}(\mathbb{L}-1)^{d+1}.$$

PROOF. The measure of the cylindrical sets can be computed as in Step 2 in 2.2. Denote $n_\tau(\mathbf{k}) + \delta + N(\mathbf{k})$ by $\bar{n}$; hence $\mu_X(V^{\mathbf{k},t}_{\bar{n},m(\mathbf{k})}) = [\pi_{\bar{n}}(V^{\mathbf{k},t}_{\bar{n},m(\mathbf{k})})]\mathbb{L}^{-(\bar{n}+1)(d+1)}$. It comes from a monomial since it is a vertex then

$$[\pi_{\bar{n}}(V^{\mathbf{k},t}_{\bar{n},m(\mathbf{k})})] = (\mathbb{L}-1)^{d+1} \mathbb{L}^{(d+1)\bar{n} - (k_1 + \ldots + k_{d+1})}.$$

∎

Hence the contribution in terms of generating functions of convex rational polyhedra is written as

$$Z_{DL}^{\Delta^t_\tau, \mathcal{W}} = T^\delta (\mathbb{L}-1)^{d+1} \mathbb{L}^{-(d+1)} \sum_{\mathbf{k} \in \Delta^t_\tau \cap \mathbb{P}^{d+1}} \mathbb{L}^{-\sigma_\omega(\mathbf{k})} T^{n_\tau(\mathbf{k}) + N(\mathbf{k})} =$$
$$= T^\delta (\mathbb{L}-1)^{d+1} \mathbb{L}^{-(d+1)} \Phi_{\Delta^t_\tau}(\mathbf{y}),$$

where $\mathbf{y} := (\mathbb{L}^{-\nu_1} T^{\alpha_1}, \ldots, \mathbb{L}^{-\nu_d} T^{\alpha_d}, \mathbb{L}^{-1} T^n)$.

For any integer linear form $\eta(v_1, \ldots, v_d, v_{d+1})$ and any $\alpha \in \mathbb{N}$ define $\Delta^\alpha_\eta := \mathbb{P}^{d+1} \cap \{\eta + \alpha = 0\}$ and $\Delta^{c,\alpha}_\eta := \mathbb{P}^{d+1} \cap \{\eta + \alpha < 0\}$. The indicator functions satisfy the identity:

$$[\Delta^{c,\alpha}_\eta] = [\Delta^{c,0}_\eta] - \sum_{0 < \alpha' \leq \alpha} [\Delta^{\alpha'}_\eta].$$

In Section 3.3 we have already used $\Delta_\eta^{c,0}$ without the superscript 0. Since $[\Delta_\tau^t \cap \mathbb{P}^{d+1}] = [\Delta_{\eta_2}^{c,\alpha_2}] - \left([\Delta_{\eta_1}^{c,\alpha_1}] + [\Delta_{\eta_1}^{\alpha_1}]\right)$ then

$$(4.8) \qquad [\Delta_\tau^t \cap \mathbb{P}^{d+1}] = \left([\Delta_{\eta_2}^{c,0}] - [\Delta_{\eta_1}^{c,0}]\right) - \left(\sum_{0<\alpha\leq\alpha_2} [\Delta_{\eta_2}^{\alpha}] - \sum_{0<\alpha<\alpha_1} [\Delta_{\eta_1}^{\alpha}] \right).$$

If the corresponding hyperplanes are parallel, then $[\Delta_\tau^t \cap \mathbb{P}^{d+1}] = \sum_{\alpha_2<\alpha<\alpha_1} [\Delta_\eta^\alpha]$ being $\eta = \tilde{\eta}_1 = \tilde{\eta}_2$.

Since η_q has integer coefficients, $\Delta_{\eta_q}^{c,0}$ is a convex (simplicial) cone with the origin as vertex. But in general the vertex of the convex rational cone $\Delta_{\eta_q}^{c,\alpha}$ is a rational $(d+1)$-tuple which may have a non-integer coordinate and then its generating function cannot be computed from the one of $\Delta_{\eta_q}^{c,0}$ with an integer translation. To avoid this problem we do the following.

Let H_q^α be the hyperplane of equation $\eta_q + \alpha = 0$, $\alpha \in \mathbb{N}$. The difference of two elements in a given H_q^α belongs to $l^q \cap \mathbb{Z}^{d+1}$. Recall that

$$l^q \cap \mathbb{Z}^{d+1} = G^q + \mathbb{Z}\mathbf{w}_1^q + \cdots + \mathbb{Z}\mathbf{w}_d^q \quad \text{and} \quad l^q \cap \mathbb{P}^{d+1} = G^q + \mathbb{N}\mathbf{w}_1^q + \cdots + \mathbb{N}\mathbf{w}_d^q,$$

where $\Delta_{\eta_q}^0 = l^q \cap \mathbb{P}^{d+1}$ and G^q is the fundamental set of $\Delta_{\eta_q}^0$:

$$G^q := \mathbb{N}^d \cap \left\{ \sum_{l=1}^{d} \mu_l \mathbf{w}_l^q \, : \, 0 < \mu_l \leq 1 \ \text{for}\, l = 1, \ldots, d \right\}.$$

We need the following lemma which we will prove later.

Lemma 4.14. *For every $\alpha \in \mathbb{N}$, there exists a subset $G^{q,\alpha} \subset H_q^\alpha$ with a bijection between G^q and $G^{q,\alpha}$ such that $H_q^\alpha \cap \mathbb{P}^{d+1} = G^{q,\alpha} + \mathbb{N}\mathbf{w}_1^q + \cdots + \mathbb{N}\mathbf{w}_d^q$.*

In particular for any $\mathbf{k} \in \Omega^{q,\alpha}$ there exists a unique $\mathbf{g} \in G^{q,\alpha}$ and $(l_1, \ldots, l_d) \in \mathbb{N}^d$ such that $\mathbf{k} = \mathbf{g} + l_1\mathbf{w}_1^q + \cdots + l_d\mathbf{w}_d^q$.

As a consequence of equation (2.1), for every $\alpha \in \mathbb{N}$, we have

$$(4.9) \qquad \Phi_{\Delta_{\eta_q}^\alpha}(\mathbf{y}) = \frac{\left(\sum_{\mathbf{g}\in G^{q,\alpha}} \mathbf{y}^{\mathbf{g}}\right)}{\prod_{l=1}^d \left(1 - \mathbf{y}^{\mathbf{w}_l^q}\right)}.$$

Assume Δ_τ^t is of type N and is the region delimited by the inequalities $\eta_q + \alpha_{q,u(q)} > 0, \eta_{q+1} + \alpha_{q+1,1} < 0$. Assume also that Δ_τ^t is different from Δ_m^t and Δ_0^t (these will be studied later on). The associated vertex $\bar{\tau}$ of $\Gamma(h)$ gives a convex cone $\Delta_{\bar{\tau}}$ delimited by the inequalities $\eta_q > 0, \eta_{q+1} < 0$. Thus the monomial $\bar{\mathbf{x}}^{\boldsymbol{\alpha}} \bar{z}^n$ appears in the sum defining h. Moreover if $\mathbf{k} \in \Delta_{\bar{\tau}}$ and $\mathbf{k}(\varphi) = \mathbf{k}$ then $\operatorname{ord}(h \circ \varphi) = n_\tau(\mathbf{k}) + N(\mathbf{k})$. As in lemma 4.13, $\mathbb{L}^{-m(\mathbf{k})}\mu_X(V^{\mathbf{k}}_{n_\tau(\mathbf{k})+N(\mathbf{k}),m(\mathbf{k})}) = \mathbb{L}^{-(d+1)}(\mathbb{L}-1)^{d+1}\mathbb{L}^{-\sigma_{\bar{\omega}}(\mathbf{k})}$.

In this case we can be more explicit in equality (4.8) because $[\Delta_{\eta_{q+1}}^{c,0}] - [\Delta_{\eta_q}^{c,0}] = [\Delta_{\bar{\tau}_q}] + [\Delta_{\bar{\gamma}_q}]$, see equation (3.10). Moreover both generating functions are evaluated at the same value $\mathbf{y} := (\mathbb{L}^{-\nu_1}T^{\alpha_1}, \ldots, \mathbb{L}^{-\nu_d}T^{\alpha_d}, \mathbb{L}^{-1}T^n)$. Thus

$$Z_{DL}^{\Delta_\tau^t,\mathcal{W}}(T) = T^\delta(\mathbb{L}-1)^{d+1}\mathbb{L}^{-(d+1)} \left(\Phi_{\Delta_{\bar{\tau}_q}}(\mathbf{y}) + \sum_{0<\alpha\leq\alpha_2} \Phi_{\Delta_{\eta_{q+1}}^\alpha}(\mathbf{y}) - \sum_{0\leq\alpha<\alpha_1} \Phi_{\Delta_{\eta_q}^\alpha}(\mathbf{y}) \right).$$

Lemma 4.15. *For each Δ_τ^t of type N its contribution to the topological zeta function is $\chi_{\text{top}}(Z_{DL}^{\Delta_\tau^t,\mathcal{W}}(h^t, \omega, \mathbb{L}^{-s})) = \chi_{\text{top}}(Z_{DL}^{\bar{\tau}}(h, \bar{\omega}, \mathbb{L}^{-s}))$.*

PROOF. The proof follows directly first from the description in equation (4.9) and the fact that in the denominator of any $\Phi_{\Delta^\alpha_\eta}(\mathbf{y})$ there are at most d factors. $\square$

REMARK 4.16. This fact will be used for both theorems 4.3 and 4.9.

The indicator function $[\Delta^t_0]$ for the first convex rational polyhedron $\Delta^t_0 = \{\eta_1 + \alpha_{1,1} < 0\}$ also satisfies the identity $[\Delta^t_0] = [\Delta^{c,0}_{\eta_1}] - \sum_{0<\alpha\leq\alpha_{1,1}}[\Delta^\alpha_{\eta_1}]$. In particular one has for this convex polyhedron a lemma similar to Lemma 4.15 too.

For the last convex rational polyhedron $\Delta^t_m = \{\eta_r + \alpha_{r,u(r)} > 0\}$, $[\Delta^t_m] = [\Delta^{c,0}_{\eta_r}] + \sum_{0\leq\alpha<\alpha_{r,u(r)}}[\Delta^\alpha_{\eta_r}]$. Thus

(4.10)

$$Z^{\Delta^t_m,\mathcal{W}}_{DL}(h^t,\omega,T) = T^\delta(\mathbb{L}-1)^{d+1}\mathbb{L}^{-(d+1)}\left(\Phi_{\Delta_{\bar{\tau}_r}}(\mathbf{y}) + \sum_{0\leq\alpha<\alpha_{r,u(r)}}\Phi_{\Delta^\alpha_{\eta_r}}(\mathbf{y})\right).$$

And again one gets for this polyhedron a lemma similar to Lemma 4.15.

If the convex rational polyhedron Δ^t_τ is of type M limited by two parallel hyperplanes, say for instance $l^t_{q,j}$ and $l^t_{q,j+1}$, then Δ^t_τ also comes from a vertex, say $t^\delta\mathbf{x}^{\boldsymbol{\alpha}}z^n$. Consider $\bar{\gamma}_q$ the corresponding compact face of $\Gamma(h)$ such that it defines l^q in the dual. One of the vertices of $\bar{\gamma}_q$ has to be $\bar{\mathbf{x}}^{\boldsymbol{\alpha}}\bar{z}^n$ because it is connected with one of the vertices of type N. We have already mentioned that $[\Delta^t_\tau \cap \mathbb{P}^{d+1}]$ is $\sum_{\alpha_{q,j+1}<\alpha<\alpha_{q,j}}[\Delta^\alpha_{\eta_q}]$. Then

$$Z^{\Delta^t_\tau,\mathcal{W}}_{DL}(h^t,\omega,T) = T^\delta(\mathbb{L}-1)^{d+1}\mathbb{L}^{-(d+1)}\left(\sum_{\alpha_{q,j+1}<\alpha<\alpha_{q,j}}\Phi_{\Delta^\alpha_{\eta_q}}(\mathbf{y})\right), \tag{4.11}$$

where $\mathbf{y} := (\mathbb{L}^{-\nu_1}T^{\alpha_1},\ldots,\mathbb{L}^{-\nu_d}T^{\alpha_d},\mathbb{L}^{-1}T^n)$. We observe that $(\alpha_1,\ldots,\alpha_d,n)$ also belongs to the closure of $\bar{\gamma}_q$. Therefore using identity (4.9) we have proved the next proposition.

PROPOSITION 4.17. *If Δ^t_τ is of type M, then $\chi_{\text{top}}(Z^{\Delta^t_\tau,\mathcal{W}}_{DL}(h^t,\omega,\mathbb{L}^{-s})) = 0$.*

We summarize the previous results and we collect also the information about the denominators of the partial *motivic* zeta functions, using the above arguments and the description of the corresponding cones given in Lemmas 3.14, 3.15 and 3.16.

PROPOSITION 4.18. *For each vertex $\bar{\tau} \in \Gamma(h)$ and each of the vertices Δ^t_τ of type M or N of $\Gamma^t(h^t)$, the functions $Z^{\Delta^t_\tau,\mathcal{W}}_{DL}(h^t,\omega,T)$ and $Z^{\bar{\tau}}_{DL}(h,\bar{\omega},T)$ belong to the subring $\mathbb{Z}[\mathbb{L},\mathbb{L}^{-1},(1-\mathbb{L}^{-1}T)^{-1},(1-\mathbb{L}^{-\sigma_\omega(\mathbf{w}^q_j)}T^{m_h(\mathbf{w}^q_j)})^{-1},(1-\mathbb{L}^{-\nu_i}T^{N_i})^{-1}][T]$, $j,i \in \{1,\ldots,d\}$, and $q = 1,\ldots,r$.*

There is a one-to-one correspondence between elements of type N and vertices of $\Gamma(h)$. If Δ^t_τ is of type N and under this bijection corresponds with $\bar{\tau} \in \Gamma(h)$ then $\chi_{\text{top}}(Z^{\Delta^t_\tau,\mathcal{W}}_{DL}(h^t,\omega,\mathbb{L}^{-s})) = \chi_{\text{top}}(Z^{\bar{\tau}}_{DL}(h,\bar{\omega},\mathbb{L}^{-s}))$.

Moreover, if Δ^t_τ is of type M then $\chi_{\text{top}}(Z^{\Delta^t_\tau,\mathcal{W}}_{DL}(h^t,\omega,\mathbb{L}^{-s})) = 0$.

PROOF OF LEMMA 4.14. Let $\mathbf{h} \in H^\alpha_q \cap \mathbb{Z}^{d+1}$, and let $\mathbf{g} \in G^q$. Write

$$\mathbf{h}+\mathbf{g} = (h_1,h_2,\ldots,h_d,h_{d+1}).$$

Using the division algorithm, for $l = 1,\ldots,d+1$, we decompose $h_l = s_l p^q_l + r_l$, where $s_l \in \mathbb{Z}$ and $0 < r_l \leq p^q_l$. Since $\mathbf{h} \in H^\alpha_q$ and $\mathbf{g} \in l^q$, then $\mathbf{h}+\mathbf{g} \in H^\alpha_q$. This implies that

$$\sum_{l=1}^{d} b_l^q r_l + \alpha = h_{d+1} n_1^q - \sum_{l=1}^{d} b_l^q s_l p_l^q = (h_{d+1} - \sum_{l=1}^{d} \bar{b}_l^q s_l) n_l^q, \text{ i.e. } h_{d+1} - \sum_{l=1}^{d} \bar{b}_l^q s_l \in \mathbb{P}.$$

Let us denote

$$\mathbf{h}_{\mathbf{g}}^1 = (r_1, ..., r_d, h_{d+1} - \sum_{l=1}^{d} \bar{b}_l^q s_l) \in H_q^\alpha \cap \mathbb{P}^{d+1}.$$

We have $\mathbf{h} + \mathbf{g} = \mathbf{h}_{\mathbf{g}}^1 + s_1 \mathbf{w}_1^q + \cdots + s_d \mathbf{w}_d^q$.

Let $\mathbf{h}' \in H_q^\alpha \cap \mathbb{P}^{d+1}$. Then $\mathbf{h}' - \mathbf{h} \in l^q \cap \mathbb{Z}^{d+1}$. There exist $\mathbf{g} \in G^q$ and $(u_1, ..., u_d) \in \mathbb{Z}^d$ such that

$$\mathbf{h}' = \mathbf{h} + \mathbf{g} + u_1 \mathbf{w}_1^q + \cdots + u_d \mathbf{w}_d^q, \text{ i.e. } \mathbf{h}' = \mathbf{h}_{\mathbf{g}}^1 + \sum_{l=1}^{d} (u_l - s_l) \mathbf{w}_l^q.$$

If $\mathbf{h}' = (h'_1, \ldots, h'_d, h'_{d+1})$ then, for $l = 1, ..., d$, we have $h'_l = r_l + (u_l - s_l) p_l^q$.

Since $h'_l \in \mathbb{P}$ and $0 < r_l \leq p_l^q$, then $(u_l - s_l) \in \mathbb{N}$. We define

$$G^{q,\alpha} := \{\mathbf{h}_{\mathbf{g}}^1 \mid \mathbf{g} \in G^q\}. \tag{4.12}$$

We have proved that $H_q^\alpha \cap \mathbb{N}_{>0}^{d+1} = G^{q,\alpha} + \mathbb{N}\mathbf{w}_1^q + \cdots + \mathbb{N}\mathbf{w}_d^q$.

It is enough to prove that G^q and $G^{q,\alpha}$ are bijective. Now let us suppose that $\mathbf{h}_{\mathbf{g}}^1 = \mathbf{h}_{\mathbf{g}'}^1$ for some $\mathbf{g}, \mathbf{g}' \in G^q$. We can write:

$$\mathbf{h} + \mathbf{g} = \mathbf{h}_{\mathbf{g}}^1 + \sum_{l=1}^{d} s_l \mathbf{w}_l^q \text{ and } \mathbf{h} + \mathbf{g}' = \mathbf{h}_{\mathbf{g}'}^1 + \sum_{l=1}^{d} s'_l \mathbf{w}_l^q.$$

We have $\mathbf{g} - \mathbf{g}' = \sum_{l=1}^{d} (s_l - s'_l) \mathbf{w}_l^q \in l^q \cap \mathbb{Z}^{d+1}$. Since the elements of this set can be written in a unique form as elements of $G^q + \mathbb{Z}\mathbf{w}_1^q + \cdots + \mathbb{Z}\mathbf{w}_d^q$ then $\mathbf{g} = \mathbf{g}'$. □

2. Edges of the Newton polytope

We now deal with edges. Consider the edges $\bar{\gamma}_1, \ldots, \bar{\gamma}_r$ of $\Gamma(h)$. Fix $q = 1, \ldots, r$. Since $f_{\boldsymbol{\lambda}_q} = \prod_{j=1}^{v(q)} (z^{n_1^q} - \bar{\beta}_j^q \mathbf{x}^{n_1^q \boldsymbol{\lambda}_q})^{m_{q,j}}$ is the quasihomogeneous part of f corresponding to this face, $[N_{\bar{\gamma}_q}] = [\mathbf{G}_{m,k}^{d+1} \cap \{f_{\boldsymbol{\lambda}_q} = 0\}] = v(q)(\mathbb{L} - 1)^d$, see Remark 3.19.

In Step 5, for each $j \in \{1, \ldots, u(q)\}$, the polynomial $\widehat{f}_{q,j}^t := \prod_{w=1}^{w(q,j)} ((z^{n_1^q})^{s_1^{q,j}} - \beta_w t^{\alpha_{q,j}} (\mathbf{x}^{n_1^q \boldsymbol{\lambda}_{\kappa,q}})^{s_1^{q,j}})^{m_{q,j,w}}$ was defined. This polynomial is associated with a region $M_t^{q,j}$ which is contained in a hyperplane parallel to l^q. Next result is trivial because under the hypothesis of the proposition there are no points with positive integer coordinates in $M_t^{q,j}$.

PROPOSITION 4.19. *If $s_1^{q,j} > 1$ then $M_t^{q,j} \cap \mathbb{N}^{d+1} = \emptyset$ and $Z_{DL}^{M_t^{q,j}, \mathcal{W}}(h^t, \omega, T) = 0$.*

From now on we will suppose $s_1^{q,j} = 1$. Because of this proposition we cannot state the equality for the topological zeta function in Theorem 4.3. For the other edges, following the notation in §2 we break the sets of arcs in A- and B-parts. We will consider,

$$\widehat{f}^t_{q,j|t=1} := \prod_{w=1}^{w(q,j)} (z^{n_1^q} - \beta_w \mathbf{x}^{n_1^q \boldsymbol{\lambda}_{\kappa,q}})^{m_{q,j,w}}.$$

REMARK 4.20. Observe that for $\widetilde{\mathcal{W}}$-quasi-ordinary and Newton compatible series, for each q we have $u(q) = 1$. The extra hypothesis $s_1^{q,j} = 1$ is fulfilled by $\widetilde{\mathcal{W}}$-quasi-ordinary and Newton compatible series and this is why equality for topological zeta functions may be obtained in Theorem 4.9.

Step 6. Edges in the A-part.

Take a convex rational cone $\Delta^t_{q,j}$ corresponding to an edge. We decompose its contribution into two disjoint parts as in the non-degenerate case. Given $\mathbf{k} \in \Delta^t_{q,j}$, we set:

$$V^{\mathbf{k},t}_{n,A,m} := \{\varphi \in \mathcal{L}_0(X) \setminus \mathcal{L}_0(S) :$$
$$\mathbf{k}(\varphi) = \mathbf{k}, \operatorname{ord}(h^t \circ \varphi) = n + \theta, \operatorname{ord}(\prod_{j=1}^{d} \bar{x}_j^{\nu_j - 1} \circ \varphi) = m, \widehat{f}^t_{q,j|t=1}(\mathbf{a}(\varphi)) \neq 0\}.$$

These semialgebraic subsets are empty unless $m = m(\mathbf{k})$ and $n = \eta_q(\mathbf{k}) + \alpha_{q,j} + N(\mathbf{k})$. In particular $[\mathbf{G}^{d+1}_{m,k} \cap \{\widehat{f}^t_{q,j|t=1} = 0\}] = w(q,j)(\mathbb{L} - 1)^d \in K_0(\mathrm{Var}_k)$, see Remark 3.19.

Let us (re)define

$$Z^{\mathcal{W}}_{q,j,A}(h^t, \omega, T) := \sum_{\mathbf{k} \in \Delta^t_{q,j}} \mathbb{L}^{-m(\mathbf{k})} \mu_X(V^{\mathbf{k},t}_{\hat{n}_q(\mathbf{k}) + \alpha_{q,j} + N(\mathbf{k}), A, m(\mathbf{k})}) T^{\eta_q(\mathbf{k}) + \alpha_{q,j} + N(\mathbf{k})}.$$

In the same way, given $\mathbf{k} \in \Delta_{\bar{\gamma}_q}$ we set:

$$V^{\mathbf{k}}_{n,A,m} := \{\varphi \in \mathcal{L}_0(X) \setminus \mathcal{L}_0(S) :$$
$$\mathbf{k}(\varphi) = \mathbf{k}, \operatorname{ord}(h \circ \varphi) = n, \operatorname{ord}(\prod_{j=1}^{d} x_j^{\nu_j - 1} \circ \varphi) = m, f_{\boldsymbol{\lambda}_{\kappa,q}}(\mathbf{a}(\varphi)) \neq 0\}.$$

Again the semialgebraic subsets are empty unless $m = m(\mathbf{k})$ and $n = \eta_q(\mathbf{k}) + N(\mathbf{k})$. We recall that

$$Z_{\bar{\gamma}_q, A}(h, \bar{\omega}, T) = \sum_{\mathbf{k} \in \Delta_{\bar{\gamma}_q}} \mathbb{L}^{-m(\mathbf{k})} \mu_X(V^{\mathbf{k}}_{\hat{n}_q(\mathbf{k}) + N(\mathbf{k}), A, m(\mathbf{k})}) T^{\eta_q(\mathbf{k}) + N(\mathbf{k})}.$$

LEMMA 4.21. *For each $q = 1, \ldots, r$, $j = 1, \ldots, u(q)$, the functions $Z^{\mathcal{W}}_{q,j,A}(h^t, \omega, T)$ and $Z_{\bar{\gamma}_q, A}(h, \bar{\omega}, T)$ belong to the subring $\mathbb{Z}[\mathbb{L}, \mathbb{L}^{-1}, (1 - \mathbb{L}^{-\sigma_\omega(\mathbf{w}_l^q)} T^{m_h(\mathbf{w}_l^q)})^{-1}][T]$, $l = 1, \ldots, d$. Moreover, the following equality holds:*

$$\sum_{j=1}^{u(q)} \chi_{\mathrm{top}}(Z^{\mathcal{W}}_{q,j,A}(h^t, \omega, \mathbb{L}^{-s})) = \chi_{\mathrm{top}}(Z_{\bar{\gamma}_q, A}(h, \bar{\omega}, \mathbb{L}^{-s})).$$

PROOF. We can follow the proofs of Theorem 2.4, Lemma 4.13 and Lemma 4.15. Since we are in the A-part of the decomposition, the measure of the sets $V^{\mathbf{k}}_{\eta_q(\mathbf{k}) + N(\mathbf{k}), A, m(\mathbf{k})}$ and $V^{\mathbf{k},t}_{\eta_q(\mathbf{k}) + \alpha_{q,j} + N(\mathbf{k}), A, m(\mathbf{k})}$ can be explicitly computed. Since h^t

and h have the same characteristic exponents then the Newton polyhedra $\Gamma^K(h^t)$ and $\Gamma(h)$ coincide and we get a sum as (4.11) but with only one $\alpha = \alpha_{q,j}$. Then

$$Z^{\mathcal{W}}_{q,j,A}(h^t,\omega,T) = \mathbb{L}^{-(d+1)}\left((\mathbb{L}-1)^{d+1} - w(q,j)(\mathbb{L}-1)^d\right) T^{\alpha_{q,j}}\Phi_{\Delta_{\eta_q}^{\alpha_{q,j}}}(\mathbf{y}), \tag{4.13}$$

where $\mathbf{y} := (\mathbb{L}^{-\nu_1}T^{\alpha_1},\ldots,\mathbb{L}^{-\nu_d}T^{\alpha_d},\mathbb{L}^{-1}T^n)$ and $(\alpha_1,\ldots,\alpha_d,n)$ is an element in the closure of $\bar{\gamma}$. In the same way

$$Z_{\bar{\gamma}_q,A}(h,\bar{\omega},T) = \mathbb{L}^{-(d+1)}\left((\mathbb{L}-1)^{d+1} - v(q)(\mathbb{L}-1)^d\right)\Phi_{\Delta_{\eta_q}^{\alpha_{q,j}}}(\mathbf{y}).$$

Since there is a bijection between the sets G^q and $G^{q,\alpha_{q,j}}$ and $v(q) = \sum w(q,j)$ for $j = 1,\ldots,u(q)$ then taking the usual Euler characteristic the lemma is proved. □

To sum up we collect the A-part in the arc decomposition as follows:

$$\begin{aligned} Z^A_{DL}(h,\bar{\omega},T) &= \sum_{\bar{\tau}\text{ vertex}} Z^{\bar{\tau}}_{DL}(h,\bar{w},T) + \sum_{q=1}^{r} Z_{\bar{\gamma}_q,A}(h,\bar{\omega},T), \\ Z^{\mathcal{W}}_A(h^t,\omega,T) &:= \sum_{\Delta^t_\tau\text{ vertex}} Z^{\Delta^t_\tau,\mathcal{W}}_{DL}(h^t,\omega,T) + \sum_{q=1}^{r}\sum_{j=1}^{u(q)} Z^{\mathcal{W}}_{q,j,A}(h^t,\omega,T). \end{aligned} \tag{4.14}$$

REMARK 4.22. Up to now we have proved two facts:

- For a $\mathcal{W}$-quasi-ordinary power series the denominators of the A-part of the motivic zeta function are controlled.
- For a $\widetilde{\mathcal{W}}$-quasi-ordinary power series which is Newton compatible, identity (3.16) also holds.

We need to replace k by $k[[t]]$ because of the computations of the B-part. Observe that in the B-part there are no vertices.

Step 7. Edges in the B-part.

Given $\mathbf{k} \in \Delta^t_{q,j}$, consider the semialgebraic subsets

$$V^{\mathbf{k},t}_{n,B,m} := \{\varphi \in \mathcal{L}_0(X)\setminus\mathcal{L}_0(S) :$$
$$\mathbf{k}(\varphi) = \mathbf{k}, \operatorname{ord}(h^t\circ\varphi) = n+\theta, \operatorname{ord}(\prod_{j=1}^{d}\bar{x}_j^{\nu_j-1}\circ\varphi) = m, \widehat{f}^t_{q,j|t=1}(\mathbf{a}(\varphi)) = 0\}.$$

If $\mathbf{k} \in \Delta_{\bar{\gamma}_q}$ we define $V^{\mathbf{k}}_{n,B,m}$ in the same way. We have to compute $Z^{\mathcal{W}}_{q,j,B}(h^t,\omega,T)$ and $Z_{\bar{\gamma}_q,B}(h,\bar{\omega},T)$ which have obvious definitions.

The sets $\Delta_{\bar{\gamma}_q}\cap\mathbb{P}^{d+1}$, resp. $\Delta^t_{q,j}\cap\mathbb{P}^{d+1}$, are the disjoint unions of the sets $\mathbf{g}+\mathbf{w}^q_1\mathbb{N}+\ldots+\mathbf{w}^q_d\mathbb{N}$ where $\mathbf{g}\in G^q$, resp. $\mathbf{g}\in G^{q,\alpha_{q,j}}$. Accordingly, we define the sets

$$V^{\bar{\gamma}_q,\mathbf{g}}_{n,B,m} := \bigcup_{\mathbf{k}\in\mathbf{g}+\mathbf{w}^q_1\mathbb{N}+\ldots+\mathbf{w}^q_d\mathbb{N}} V^{\mathbf{k}}_{n,B,m}, \qquad \mathbf{g}\in G^q,$$

and

$$V^{q,j,\mathbf{g},t}_{n,B,m} := \bigcup_{\mathbf{k}\in\mathbf{g}+\mathbf{w}^q_1\mathbb{N}+\ldots+\mathbf{w}^q_d\mathbb{N}} V^{\mathbf{k},t}_{n,B,m}, \qquad \mathbf{g}\in G^{q,\alpha_{q,j}}.$$

Let $\varphi \in V^{\bar{\gamma}_q,\mathbf{g}}_{n,B,m}$, with $\mathbf{g}\in G^q$; this arc is related to exactly one Newton component of f, see Section 3.4. There is a unique $\bar{j} = 1,\ldots,v(q)$ such that $\mathbf{a}(\varphi)\in\mathbf{G}^{d+1}_{m,k}$ belongs to the zero locus $V^q_{\bar{j}}\subset\mathbb{A}^{d+1}_k$ of the quasi-ordinary polynomial $f_{q,j} = z^{n^q_1} - \bar{\beta}^q_{\bar{j}}\mathbf{x}^{n^q_1\boldsymbol{\lambda}_q}$.

REMARK 4.23. Let J_q be the subset of $\{1,\ldots,d\}$ of the non-zero coordinates of $\boldsymbol{\lambda}_q$. It may happen that $J_q \subsetneq \{1,\ldots,d\}$. For the sake of simplicity we suppose that there is equality and we will point out where the non-equality may appear.

We decompose the set $V_{n,B,m}^{\bar{\gamma}_q,\mathbf{g}}$ in $v(q)$ disjoint sets according to this property:

$$V_{n,B,m}^{\bar{\gamma}_q,\mathbf{g}} = \bigcup_{\bar{j}=1}^{v(q)} V_{n,B,m}^{\bar{\gamma}_q,\mathbf{g},\bar{j}}.$$

In the same way the sets $V_{n,B,m}^{t,q,j,\mathbf{g},w}$, $w = 1,\ldots,w(q,j)$, are defined. It is also possible to decompose

$$V_{n,B,m}^{\bar{\gamma}_q,\mathbf{g},\bar{j}} = \bigcup_{\mathbf{r}:=(r_1,\ldots,r_d)\in\mathbb{N}^d} V_{n,B,m}^{\bar{\gamma}_q,\mathbf{g},\bar{j},\mathbf{r}},$$

where $V_{n,B,m}^{\bar{\gamma}_q,\mathbf{g},\bar{j},\mathbf{r}} := \{\varphi \in V_{n,B,m}^{\bar{\gamma}_q,\mathbf{g},\bar{j}} : \mathbf{k}(\varphi) = \mathbf{g} + r_1\mathbf{w}_1^q + \ldots + r_d\mathbf{w}_d^q\}$, and consider in the same way $V_{n,B,m}^{t,q,j,\mathbf{g},w,\mathbf{r}}$. Define

$$Z^{\bar{\gamma}_q,\mathbf{g},\bar{j},\mathbf{r}}(T) := \sum_{n\geq 1}\left(\sum_{m\geq 1} \mathbb{L}^{-m}\mu_X(V_{n,B,m}^{\bar{\gamma}_q,\mathbf{g},\bar{j},\mathbf{r}})\right)T^n,$$

and also $Z_{q,j}^{\mathcal{W},\mathbf{g},w,\mathbf{r}}(T)$. We define $Z^{\bar{\gamma}_q,\mathbf{g},\bar{j}}(T) := \sum_{\mathbf{r}\in\mathbb{N}^d} Z^{\bar{\gamma}_q,\mathbf{g},\bar{j},\mathbf{r}}(T)$ and the corresponding $Z_{q,j}^{\mathcal{W},\mathbf{g},w}(T)$; then we have the decompositions:

$$(4.15)\qquad \begin{aligned} Z_{DL}^{\mathcal{W}}(h^t,\omega,T) &= Z_A^{\mathcal{W}}(h^t,\omega,T) + \sum_{q=1}^{r}\sum_{j=1}^{u(q)}\sum_{\mathbf{g}\in G^{q,\alpha_{q,j}}}\sum_{w=1}^{w(q,j)} Z_{q,j}^{\mathcal{W},\mathbf{g},w}(T) \\ Z_{DL}(h,\bar{\omega},T) &= Z_{DL}^{A}(h,\bar{\omega},T) + \sum_{q=1}^{r}\sum_{\mathbf{g}\in G^q}\sum_{\bar{j}=1}^{v(q)} Z^{\bar{\gamma}_q,\mathbf{g},\bar{j}}(T), \end{aligned}$$

where $v(q) = \sum_{j=1}^{u(q)} w(q,j)$. Using Proposition 4.18 and Lemma 4.21 we have proved that $Z_A^{\mathcal{W}}(h^t,\omega,T)$ satisfies Theorems 4.3 and 4.9 in their respective cases. Then we have to prove them for arcs in the B-part of the decomposition. In particular we will describe in the following steps the computations in the identities of (4.15).

Step 8. Computations for h and $\mathbf{g} = \mathbf{g}^1 := \mathbf{w}_1^q + \ldots + \mathbf{w}_d^q \in G^q$, and $\bar{j} = 1,\ldots,v(q)$.

We use the notation of Section 3.4. To compute $Z^{\bar{\gamma}_q,\mathbf{g}^1,\bar{j},\mathbf{r}}(T)$, for $\mathbf{r}\in\mathbb{N}^d$, then $\mathbf{k} = (r_1+1)\mathbf{w}_1^q + \ldots + (r_d+1)\mathbf{w}_d^q$ has coordinates $k_l = (r_l+1)p_l^q$ with $l = 1,\ldots,d$, and $k_{d+1} = \sum_{l=1}^d (r_l+1)\bar{b}_l^q$. If φ satisfies that $\mathbf{k}(\varphi) = \mathbf{k}$ then $m_{\mathbf{k}} := \operatorname{ord}(\prod_{j=1}^d \bar{x}_j^{\nu_j-1}\circ\varphi) = \sum_{l=1}^d (r_l+1)p_l^q(\nu_l-1)$. Therefore $m_{\mathbf{k}}$ is the unique value of m for which $V_{n,B,m}^{\bar{\gamma}_q,\mathbf{g}^1,\bar{j},\mathbf{r}}$ may be non-empty. Hence

$$Z^{\bar{\gamma}_q,\mathbf{g}^1,\bar{j},\mathbf{r}}(T) = \sum_{n\geq 1} \mathbb{L}^{-m_{\mathbf{k}}}\mu_X(V_{n,B,m_{\mathbf{k}}}^{\bar{\gamma}_q,\mathbf{g}^1,\bar{j},\mathbf{r}})T^n.$$

Fix the parametrization $\pi_{\bar{j}}^q : \mathbb{A}_k^d \to V_{\bar{j}}^q$ as in Section 3.4. Take $\varphi \in V_{n,B,m_{\mathbf{k}}}^{\bar{\gamma}_q,\mathbf{g}^1,\bar{j},\mathbf{r}}$ and the unique $\mathbf{s}^0 \in \mathbf{G}_{m,k}^d$ such that $\pi_{\bar{j}}^q(\mathbf{s}^0) = \mathbf{a}(\varphi)$. The idea is to lift φ to an affine space $Y := \mathbb{A}_k^{d+1}$ with coordinates $(\mathbf{y}, z_1)$ using the *Newton map* $\pi_{q,\bar{j}}$ associated with $f_{\bar{j}}^q$.

This Newton map defines a $k[t]$-morphism $\pi_{q,\bar{j}} : \mathcal{L}(Y) \to \mathcal{L}(X)$ (in fact nothing is done on the variable t). We can apply to $\pi_{q,\bar{j}}$ the change of variables formula, see Theorem 1.2.

For $l = 1, \ldots, d$, the l-th component of $\boldsymbol{\varphi}$ is $\varphi_l(t) = t^{(r_l+1)p_l^q} v_l(t)$, where $v_l(t) \in k[[t]]$ such that $v_l(0) = (s_l^0)^{p_l^q}$. There exists a unique $w_l(t) \in k[[t]]$ such that $w_l(0) = s_l^0$ and $w_l(t)^{p_l^q} = v_l(t)$. If $\psi_l(t) := t^{(r_l+1)} w_l(t)$ then

$$\varphi_{d+1}(t) = t^{(r_1+1)\bar{b}_1^q + \ldots + (r_d+1)\bar{b}_d^q} v_{d+1}(t) \text{ and } v_{d+1}(0) = \prod_{l=1}^{d} (s_l^0)^{\bar{b}_l^q}.$$

The identity

$$\begin{aligned} t^{(r_1+1)\bar{b}_1^q + \ldots + (r_d+1)\bar{b}_d^q} v_{d+1}(t) =& (\psi_{d+1}(t) + \bar{\beta}_{\bar{j}}^q) \psi_1(t)^{\bar{b}_1^q} \ldots \psi_d(t)^{\bar{b}_d^q} = \\ =& (\psi_{d+1}(t) + \bar{\beta}_{\bar{j}}^q) t^{(r_1+1)\bar{b}_1^q + \ldots + (r_d+1)\bar{b}_d^q} w_1(t)^{\bar{b}_1^q} \ldots w_d(t)^{\bar{b}_d^q} \end{aligned}$$

completely determines a series $\psi_{d+1}(t)$ such that $\psi_{d+1}(0) = 0$. So there is a unique $\boldsymbol{\psi} := (\psi_1, \ldots, \psi_d, \psi_{d+1}) \in \mathcal{L}_0(Y)$ such that $\pi_{q,\bar{j}}(\boldsymbol{\psi}) = \boldsymbol{\varphi}$.

Moreover, if $\bar{h}_{q,\bar{j}} := h \circ \pi_{q,\bar{j}}$, we denote by $B_{n,m_{\mathbf{k}}}^{\mathbf{r}}$ the set of arcs $\boldsymbol{\psi} \in \mathcal{L}_0(Y)$ such that

- $\operatorname{ord}(\bar{h}_{q,\bar{j}} \circ \boldsymbol{\psi}) = n$,
- $\operatorname{ord}(\prod_{j=1}^{d} x_j^{\nu_j - 1} \circ \pi_{q,\bar{j}} \circ \boldsymbol{\psi}) = m_{\mathbf{k}} = \sum_{l=1}^{d} (r_l+1) p_l^q (\nu_l - 1)$,
- if $\mathbf{k}(\boldsymbol{\psi}) = (\tilde{k}_1, \ldots, \tilde{k}_d, \tilde{k}_{d+1})$ then $\tilde{k}_l = r_l + 1$, for $l = 1, \ldots, d$.

Then $\pi_{q,\bar{j}}$ defines a bijection between $B_{n,m_{\mathbf{k}}}^{\mathbf{r}}$ and $V_{n,B,m_{\mathbf{k}}}^{\bar{\gamma}_q,\mathbf{g}^1,\bar{j},\mathbf{r}}$. The order of the Jacobian of the $k[t]$-morphism $\pi_{q,\bar{j}}$ is constant on $B_{n,m_{\mathbf{k}}}^{\mathbf{r}}$ and equals to:

$$\operatorname{ord}_t(\mathcal{J}_{\pi_{q,\bar{j}}}) = \sum_{l=1}^{d} (r_l+1)(p_l^q + \bar{b}_l^q - 1).$$

The set $B_{n,m_{\mathbf{k}}}^{\mathbf{r}}$ is strongly measurable in $\mathcal{L}(Y)$. Hence $V_{n,B,m_{\mathbf{k}}}^{\bar{\gamma}_q,\mathbf{g}^1,\bar{j},\mathbf{r}}$, its bijective image by $\pi_{q,\bar{j}}$, is also strongly measurable in $\mathcal{L}(X)$, [**18**, Theorem A.8]. To measure $V_{n,B,m_{\mathbf{k}}}^{\bar{\gamma}_q,\mathbf{g}^1,\bar{j},\mathbf{r}}$ the change variables formula gives

$$\mu_X(V_{n,B,m_{\mathbf{k}}}^{\bar{\gamma}_q,\mathbf{g}^1,\bar{j},\mathbf{r}}) = \int_{V_{n,B,m_{\mathbf{k}}}^{\bar{\gamma}_q,\mathbf{g}^1,\bar{j},\mathbf{r}}} d\mu_X = \int_{B_{n,m_{\mathbf{k}}}^{\mathbf{r}}} \mathbb{L}^{-\operatorname{ord}_t \mathcal{J}_{\pi_{q,\bar{j}}}(y)} d\mu_Y.$$

If $\widehat{\omega}_1$ is the pullback by $\pi_{q,\bar{j}}$ of the differential form $d\bar{x}_1 \wedge \cdots \wedge d\bar{x}_d \wedge d\bar{z}$ then the measure $\mu_X(V_{n,B,m_{\mathbf{k}}}^{\bar{\gamma}_q,\mathbf{g}^1,\bar{j},\mathbf{r}})$ is equal to

$$\sum_{k \in \mathbb{P}} \mathbb{L}^{-k} \mu_Y(B_{n,m_{\mathbf{k}}}^{\mathbf{r}} \cap \{\operatorname{ord}_t(\widehat{\omega}_1) = k\}) = \mathbb{L}^{-\sum_{l=1}^{d}(r_l+1)(p_l^q + \bar{b}_l^q - 1)} \mu_Y(B_{n,m_{\mathbf{k}}}^{\mathbf{r}}).$$

Putting all these terms together

$$\begin{aligned} Z^{\bar{\gamma}_q,\mathbf{g}^1,\bar{j},\mathbf{r}}(T) &= \sum_{n \geq 1} \mathbb{L}^{-m_{\mathbf{k}} - \sum_{l=1}^{d}(r_l+1)(p_l^q + \bar{b}_l^q - 1)} \mu_Y(B_{n,m_{\mathbf{k}}}^{\mathbf{r}}) T^n \\ &= \sum_{n \geq 1} \mathbb{L}^{-\sum_{l=1}^{d}(r_l+1)(p_l^q \nu_l + \bar{b}_l^q - 1)} \mu_Y(B_{n,m_{\mathbf{k}}}^{\mathbf{r}}) T^n. \end{aligned}$$

The pull-back $\omega_{q,\bar{j}} := \bar{\omega}\circ\pi_{q,\bar{j}}$ of $\bar{\omega}$ under $\pi_{q,\bar{j}}$ is $\omega_{q,\bar{j}} = \prod_{l=1}^{d} \bar{y_l}^{(p_l^q\nu_l+\bar{b}_l^q-1)} d\bar{y}_1\wedge d\bar{y}_d\wedge dz_1$. In particular $\mathrm{ord}_t(\prod_{l=1}^{d} \bar{y_l}^{(p_l^q\nu_l+\bar{b}_l^q-1)} \circ \psi) = \sum_{l=1}^{d}(r_l+1)(p_l^q\nu_l+\bar{b}_l^q-1)$ for every arc $\psi \in B^r_{n,m_{\mathbf{k}}}$. If $\tilde{m_{\mathbf{k}}} := \mathrm{ord}_t(\prod_{l=1}^{d} \bar{y_l}^{(p_l^q\nu_l+\bar{b}_l^q-1)} \circ \psi)$ then

$$Z^{\bar{\gamma}_q,\mathbf{g}^1,\bar{j},\mathbf{r}}(T) = \sum_{n\geq 1} \mathbb{L}^{-\tilde{m_{\mathbf{k}}}}\mu_Y(B^{\mathbf{r}}_{n,m_{\mathbf{k}}})T^n.$$

The sets $B^{\mathbf{r}}_{n,m}$ are empty if $m \neq \tilde{m_{\mathbf{k}}}$ so for all $\bar{j} = 1,\ldots,v(q)$

$$Z^{\bar{\gamma}_q,\mathbf{g}^1,\bar{j}}(T) = \sum_{\mathbf{r}\in\mathbb{N}^d}\sum_{n\geq 1} \mathbb{L}^{-\tilde{m_{\mathbf{k}}}}\mu_Y(B^{\mathbf{r}}_{n,m_{\mathbf{k}}})T^n = Z_{DL}(\bar{h}_{q,\bar{j}},\omega_{q,\bar{j}},T). \tag{4.16}$$

REMARK 4.24. If h has $\mathbf{ev}(h) = 1$, no more steps are needed to get a formula for $Z_{DL}(h,\omega,T)$, see Corollary 5.2. This happens because each fundamental set G^q has only one element.

Step 9. Computations for h^t, $\mathbf{g} \in G^{q,\alpha_{q,j}}$ and $w = 1,\ldots,w(q,j)$.

By the proof of Lemma 4.14, and more precisely by identity (4.12), one can write $\mathbf{g} = (\mathbf{0}, \frac{\alpha_{q,j}}{n_1^q}) + \sum_{l=1}^{d} \mu_l^{\mathbf{g}}\mathbf{w}_l^q$. Thus

$$\mathbf{g} = (\mu_1^{\mathbf{g}}p_1^q,\ldots,\mu_d^{\mathbf{g}}p_d^q, \frac{\alpha_q}{n_1^q} + \sum_{l=1}^{d}\mu_l^{\mathbf{g}}\bar{b}_l^q) \in \mathbb{P}^{d+1}.$$

By definition of $G^{q,\alpha_{q,j}}$, $0 < \mu_l^{\mathbf{g}} \leq 1$ is rational. Choose $v_1,\ldots,v_d \in \mathbb{P}$ such that $\mu_l^{\mathbf{g}} = \frac{v_l}{p_l^q}$, $l = 1,\ldots,d$ and let $c^{\mathbf{g}} \in \mathbb{P}$ the $(d+1)^{\text{th}}$-coordinate of $\mathbf{g}$.

Fix $\mathbf{r} \in \mathbb{N}^d$. Let $\varphi \in V^{t,q,j,\mathbf{g},w,\mathbf{r}}_{n,B,m} = \{\varphi \in V^{t,q,j,\mathbf{g}}_{n,B,m} : \mathbf{k}(\varphi) = \mathbf{g}+r_1\mathbf{w}_1+\ldots+r_d\mathbf{w}_d\}$ be an arc and denote:

- $\mathbf{k} := \mathbf{k}(\varphi) = \left((\mu_1^{\mathbf{g}}+r_1)p_1^q,\ldots,(\mu_d^{\mathbf{g}}+r_d)p_d^q, \frac{\alpha_{q,j}}{n_1^q} + \sum_{l=1}^{d}(\mu_l^{\mathbf{g}}+r_l)\bar{b}_l^q\right)$.

- $\mathbf{a} := \mathbf{a}(\varphi) \in \mathbf{G}^{d+1}_{m,k} \cap \{\widehat{f}^{t,w}_{q,j} = 0\}$, where $\widehat{f}^{t,w}_{q,j} = z^{n_1^q} - \beta_w\mathbf{x}^{n_1^q\boldsymbol{\lambda}_{\kappa_q}}$. Consider

$\mathbf{a} = \pi_j^{q,w}(s^0)$, $s^0 \in \mathbf{G}^d_{m,k}$ with the fixed parametrization of $\{\widehat{f}^{t,w}_{q,j} = 0\}$ induced by $\pi_j^{q,w} : \mathbb{A}^d_k \to \mathbf{G}^{d+1}_{m,k} \cap \{\widehat{f}^{t,w}_{q,j} = 0\}$: $(s_1,\ldots,s_d) \mapsto (s_1^{p_1^q},\ldots,s_d^{p_d^q},\alpha_w s_1^{\bar{b}_1^q}\ldots s_d^{\bar{b}_d^q})$ where $(\alpha_w)^{n_1^q} = \beta_w$.

For each φ with $\mathbf{k}(\varphi) = \mathbf{k}$ then $m_{\mathbf{k}} := \mathrm{ord}(\prod_{j=1}^{d} x_j^{\nu_j-1}\circ\varphi) = \sum_{l=1}^{d}(r_l+\mu_l^{\mathbf{g}})p_l^q(\nu_l-1)$. Since $m_{\mathbf{k}}$ is the unique value of m for which $V^{t,q,j,\mathbf{g},w,\mathbf{r}}_{n,B,m}$ may be non-empty then

$$Z^{\mathcal{W},\mathbf{g},w}_{q,j}(T) = \sum_{r\in\mathbb{N}^d} Z^{\mathcal{W},\mathbf{g},w,\mathbf{r}}_{q,j}(T) = \sum_{r\in\mathbb{N}^d} \mathbb{L}^{-m_{\mathbf{k}}}\mu_X(V^{t,q,j,\mathbf{g},w,\mathbf{r}}_{n,B,m_{\mathbf{k}}})T^n.$$

Let $Y = \mathbb{A}^{d+1}_k$ and let $\pi_{\mathbf{g}} : \mathcal{L}(Y) \to \mathcal{L}(X)$ be the $k[t]$-morphism defined by $x_l = t^{v_l}y_l$, for every $l = 1,\ldots,d$, and $z = t^{c^{\mathbf{g}}}z_1$.

LEMMA 4.25. *Under the above conditions:*

(1) *The map* $\pi_{\mathbf{g}}$ *defines a* $k[t]$*-morphism.*
(2) *The Jacobian of* $\pi_{\mathbf{g}}$ *has constant order* $\nu_{\mathbf{g}} := \operatorname{ord}_t(J_{\pi_{\mathbf{g}}}) = \sum_{l=1}^{d} v_l + c^{\mathbf{g}}$ *on the arcs of* $\mathcal{L}(Y)$.
(3) *The* K*-quasi-ordinary function* $h^{\{t\}} := h^t \circ \pi_{\mathbf{g}}$ *has the same characteristic exponents (and monomials) as* h^t*, and consequently the same as* h *(this follows directly from Remark 3.35).*

Recall that $h^t = t^{\theta} x_1^{N_1} x_2^{N_2} \dots x_d^{N_d} f^t(\mathbf{x}, z) u^t(\mathbf{x}, z)$ but we are assuming for simplicity that $\theta = 0$. Then $h^{\{t\}} = t^{\tilde{A}_{\mathbf{g}}} y_1^{N_1} \dots y_d^{N_d} f^{\{t\}}(\mathbf{y}, z_1) u^{\{t\}}(\mathbf{y}, z_1)$.

Fix $\mathbf{r} \in \mathbb{N}^d$, and take an arc $\boldsymbol{\varphi} \in V_{n,B,m_{\mathbf{k}}}^{t,q,j,\mathbf{g},w,\mathbf{r}}$. There exists a unique arc $\boldsymbol{\psi} \in \mathcal{L}(Y)$ such that $\boldsymbol{\varphi} = \pi_{\mathbf{g}} \circ \boldsymbol{\psi}$. Properties of $\boldsymbol{\psi}$ will depend on $\mathbf{r}$:

(2.a) $\mathbf{r} \in \mathbb{P}^d$; in this case $\boldsymbol{\psi} \in \mathcal{L}_0(Y)$, $\mathbf{a}(\boldsymbol{\psi}) = \mathbf{a}$ and

$$\mathbf{k}(\boldsymbol{\psi}) = \left(r_1 p_1^q, \dots, r_d p_d^q, \sum_{l=1}^{d} r_l \bar{b}_l^q \right).$$

(2.b) $\mathbf{r} \in \mathbb{N}^d \setminus \mathbb{P}^d$; in this case $\boldsymbol{\psi} \in \mathcal{L}_{(\mathbf{y},0)}(Y)$. More precisely; let $J_{\mathbf{r}} \subsetneq \{1, \dots, d\}$ be the set of indices such that $r_l = 0 \Leftrightarrow l \in J_{\mathbf{r}}$; we observe that $\emptyset \neq J_{\mathbf{r}} \subsetneq \{1, \dots, d\}$. The point $\mathbf{y}$ corresponds to $(y_1, \dots, y_d)$ where $y_l = (s_l^0)^{p_l^q}$ if $l \in J_{\mathbf{r}}$ and $y_l = 0$ otherwise. We include in this case the case where $\mathbf{r} = \mathbf{0} \in \mathbb{N}^d$ and $J = \{1, \dots, d\}$.

REMARK 4.26. Note that in fact $J \subset J_q$, see Remark 4.23.

Decompose $Z_{q,j}^{\mathcal{W},\mathbf{g},w}(T) = Z_{1,q,j}^{\mathcal{W},\mathbf{g},w}(T) + \sum_{\emptyset \subsetneq J \subset \{1,\dots,d\}} Z_{J,q,j}^{\mathcal{W},\mathbf{g},w}(T)$, where

$$Z_{1,q,j}^{\mathcal{W},\mathbf{g},w}(T) := \sum_{\mathbf{r} \in \mathbb{P}^d} \sum_{n \geq 1} \mathbb{L}^{-m_{\mathbf{k}}} \mu_X(V_{n,B,m_{\mathbf{k}}}^{t,q,j,\mathbf{g},w,\mathbf{r}}) T^n$$

and

$$Z_{J,q,j}^{\mathcal{W},\mathbf{g},w}(T) := \sum_{\mathbf{r} \in \mathbb{N}^d : J = J_{\mathbf{r}}} \sum_{n \geq 1} \mathbb{L}^{-m_{\mathbf{k}}} \mu_X(V_{n,B,m_{\mathbf{k}}}^{t,q,j,\mathbf{g},w,\mathbf{r}}) T^n.$$

Define $\omega_{\mathbf{g}}^{\{t\}} := \pi_{\mathbf{g}}^* \omega$ and for $\mathbf{r} \in \mathbb{P}^d$ denote by $W_{n,B,m_{\mathbf{k}}}^{t,q,j,\mathbf{g},w,\mathbf{r}}$ the semialgebraic subset of arcs $\boldsymbol{\psi} \in \mathcal{L}_0(Y)$ such that:

- $\operatorname{ord}(h^{\{t\}} \circ \boldsymbol{\psi}) = n + \theta$ (but $\theta = 0$ for simplicity). If $n' := n - \tilde{A}_{\mathbf{g}}$ then $\operatorname{ord}(h^{\{t\}} \circ \boldsymbol{\psi}) = n' + \tilde{A}_{\mathbf{g}}$.
- $\mathbf{k}(\boldsymbol{\psi}) = (r_1 p_1^q, \dots, r_d p_d^q, \sum_{l=1}^{d} r_l \bar{b}_l^q)$.
- $\operatorname{ord}(\prod_{j=1}^{d} x_j^{\nu_j - 1} \circ \pi_{\mathbf{g}} \circ \boldsymbol{\psi}) = m_{\mathbf{k}}$. The differential form $\omega_{\mathbf{g}}^{\{t\}}$ is defined as $t^{A_{\mathbf{g}}} \omega_{\mathbf{g}}$ where $\omega_{\mathbf{g}} := \prod_{l=1}^{d} y_l^{\nu_l - 1} d\mathbf{y} \wedge dz_1$ and $A_{\mathbf{g}} := \sum_{l=1}^{d} v_l(\nu_l - 1)$. If $\tilde{m}_{\mathbf{k}} := \operatorname{ord}(\prod_{l=1}^{d} y_l^{\nu_l - 1} \circ \boldsymbol{\psi})$ then $m_{\mathbf{k}} = \tilde{m}_{\mathbf{k}} + A_{\mathbf{g}}$ and

$$\nu_{\mathbf{g}} + A_{\mathbf{g}} = \frac{\alpha_q}{n_1^q} + \sum_{l=1}^{d} \mu_l^{\mathbf{g}} (p_l^q \nu_l + \bar{b}_l^q).$$

The map $\pi_{\mathbf{g}}$ induces a bijection between the sets $V_{n,B,m_{\mathbf{k}}}^{t,q,j,\mathbf{g},w,\mathbf{r}}$ and $W_{n,B,m_{\mathbf{k}}}^{t,q,j,\mathbf{g},w,\mathbf{r}}$. Since $\mu_X(V_{n,B,m_{\mathbf{k}}}^{t,q,j,\mathbf{g},w,\mathbf{r}}) = \mathbb{L}^{-\nu_{\mathbf{g}}}\mu_Y(W_{n,B,m_{\mathbf{k}}}^{t,q,j,\mathbf{g},w,\mathbf{r}})$, applying the change of variables formula to $\pi_{\mathbf{g}}$ we have

$$\begin{aligned} Z_{1,q,j}^{\mathcal{W},\mathbf{g},w}(T) &= \sum_{\mathbf{r}\in\mathbb{P}^d}\sum_{n\geq 1}\mathbb{L}^{-m_{\mathbf{k}}-\nu_{\mathbf{g}}}\mu_Y(W_{n,B,m_{\mathbf{k}}}^{t,q,j,\mathbf{g},w,\mathbf{r}})T^n \\ &= \sum_{\mathbf{r}\in\mathbb{P}^d}\sum_{n\geq 1}\mathbb{L}^{-\nu_{\mathbf{g}}-A_{\mathbf{g}}-\mathrm{ord}_t(\omega_{\mathbf{g}})}\mu_Y(W_{n,B,m_{\mathbf{k}}}^{t,q,j,\mathbf{g},w,\mathbf{r}})T^n \\ &= \mathbb{L}^{-(\nu_{\mathbf{g}}+A_{\mathbf{g}})}\sum_{\mathbf{r}\in\mathbb{P}^d}\sum_{n\geq 1}\mathbb{L}^{-\tilde{m}_{\mathbf{k}}}\mu_Y(W_{n,B,\tilde{m}_{\mathbf{k}}}^{t,q,j,\mathbf{g}^1,w,\mathbf{r}})T^n \\ &= \mathbb{L}^{-(\nu_{\mathbf{g}}+A_{\mathbf{g}})}\sum_{\mathbf{r}\in\mathbb{P}^d}T^{\tilde{A}_{\mathbf{g}}}\sum_{n'\geq 1}\mathbb{L}^{-\tilde{m}_{\mathbf{k}}}\mu_Y(W_{n,B,\tilde{m}_{\mathbf{k}}}^{t,q,j,\mathbf{g}^1,w,\mathbf{r}})T^{n'}. \end{aligned} \tag{4.17}$$

Consider the following Newton map associated with the Newton component of h^t (as a K-quasi-ordinary power series) we are dealing with. Its Newton polyhedron is defined by $z^{n_1^q} - \boldsymbol{\beta}_{j,w}^q(t)x_1^{b_1^q}\dots x_d^{b_d^q}$, where $\boldsymbol{\beta}_{j,w}^q(t) = \beta_w t^{\alpha_{q,j}}$+higher order terms $\in k[[t]]$ and $\beta_w \in \mathbf{G}_{m,k}$. We can write $\boldsymbol{\beta}_{j,w}^q(t) = t^{\alpha_{q,j}}\alpha(t)$ with $\alpha(t) = (\alpha_w + \dots)^{n_1^q} \in k[[t]]$ because $(\alpha_w)^{n_1^q} = \beta_w$.

Let $\pi_{j,t}^{q,w}$ be the $k[[t]]$-morphism associated to the transformation $y_l = \tilde{y}_l^{p_l^q}$ for all $i = 1,\dots,d$ and $z_1 = (z_2 + \alpha(t))\prod_{l=1}^{d}\tilde{y}_l^{\bar{b}_l^q}$. Recall the change of variables formula is also applied here, see Remark 1.4.

Taking into account the pull-back of the discriminants and applying the same proof as in Step 8, the following proposition follows.

PROPOSITION 4.27. *Let $h_{q,j,\mathbf{g},w}^t := h^{\{t\}} \circ \pi_{j,t}^{q,w} = h^t \circ \pi_{\mathbf{g}} \circ \pi_{j,t}^{q,w}$ and $\tilde{\omega}_{q,j,\mathbf{g},w} = (\pi_{j,t}^{q,w})^*\omega_{\mathbf{g}}$. Then:*

(1) *The series $h_{q,j,\mathbf{g},w}^t$ is $\mathcal{W}$-quasi-ordinary and its depth is strictly less than that of h^t. We have*

$$Z_{1,q,j}^{\mathcal{W},\mathbf{g},w}(T) = \mathbb{L}^{-(\nu_{\mathbf{g}}+A_{\mathbf{g}})}Z_{DL}^{\mathcal{W}}(h_{q,j,\mathbf{g},w}^t, \tilde{\omega}_{q,j,\mathbf{g},w}, T). \tag{4.18}$$

(2) *If h^t is $\widetilde{\mathcal{W}}$-quasi-ordinary and Newton compatible then there exists a weight $\widetilde{\mathcal{W}}'$ such that $h_{q,j,\mathbf{g},w}^t$ is $\widetilde{\mathcal{W}}'$-quasi-ordinary and Newton compatible.*
(3) *The series $h_{q,j,\mathbf{g},w}^t$ is J_q-bounded, see Remark 4.23.*

REMARK 4.28. In the identity (4.18) we recall the definition (4.4) of the zeta function $Z_{DL}^{\mathcal{W}}(h_{q,j,\mathbf{g},w}^t, \tilde{\omega}_{q,j,\mathbf{g},w}, T)$:

$$T^{\tilde{A}_{\mathbf{g}}}\sum_{n'\in\mathbb{N}}\left(\sum_{m\in\mathbb{N}}\mathbb{L}^{-m}\mu(V_{n,m}^t)\right)T^{n'}.$$

PROOF. The only non trivial facts are the two last ones. It is easy to prove from the Newton map that $h_{q,j,\mathbf{g},w}^t$ is $\widetilde{\mathcal{W}}'$-quasi-ordinary, where

$$\widetilde{\mathcal{W}}' := (pn_1^q; (pv_1 + g_1)c_1^q, \dots, (pv_d + g_d)c_d^q).$$

This is due to the following equalities:

$$\alpha_{q,j}=\frac{1}{p}\sum_{l=1}^{d}b_l^q g_l,\quad c^{\mathbf{g}}=\frac{\alpha_{q,j}}{n_1^q}+\sum_{l=1}^{d}\frac{v_l}{p_l^q}\bar{b}_l^q=\frac{\alpha_{q,j}}{n_1^q}+\sum_{l=1}^{d}\frac{v_l}{n_1^q}b_l^q=\frac{1}{n_1^q}\sum_{l=1}^{d}(v_l+\frac{g_l}{p})b_l^q.$$

The $\widetilde{\mathcal{W}}'$ depends only on $\widetilde{\mathcal{W}}$ and on the Newton map. This implies that we can choose a *generic* $\widetilde{\mathcal{W}}$-quasi-ordinary series $\tilde{h}^t$ with the same characteristic monomials and then Newton compatible. Then, $\tilde{h}^t_{q,j,\mathbf{g},w}$ is also $\widetilde{\mathcal{W}}'$-quasi-ordinary. If we pass to good coordinates the weight $\widetilde{\mathcal{W}}'$ remains and both power series have the same Newton polyhedron. For $\tilde{h}^t_{q,j,\mathbf{g},w}$ all the coefficients in its restriction to the Newton polyhedron will be non zero and it is easily seen that it implies that $\tilde{h}^t_{q,j,\mathbf{g},w}$ is Newton compatible and it is also the case for $h^t_{q,j,\mathbf{g},w}$.

The fact that $h^t_{q,j,\mathbf{g},w}$ is J_q-bounded follows from the behavior of the t-morphism $\pi_{\mathbf{g}}$; note that each variable of J_q is replaced by this variable times a power of t; then for a fixed power of t only a finite number of monomials in t and the variables of J_q can contribute. □

Consider now arcs in Case (2.b). Let $\emptyset \subsetneq J\subset\{1,\dots,d\}$; in fact, from Remark 4.26, we are only interested in the case $\emptyset\subsetneq J\subset J_q$. Let us fix a point $(\mathbf{y},0)$ corresponding to the subset J. Since $h^t_{q,j,\mathbf{g},w}$ is $\mathcal{W}$-quasi-ordinary J-bounded power series the germ of $h^t_{q,j,\mathbf{g},w}$ at $(\mathbf{y},0)$ is also a $\mathcal{W}$-quasi-ordinary power series, see Remark 4.2. We can indeed evaluate at $(\mathbf{y},0)$ because the coefficients of its monomials in t belong to $k[\widetilde{\mathbf{y}},z]$. Since $h^t_{q,j,\mathbf{g},w}$ is $\mathcal{W}$-quasi-ordinary its discriminant is computed from the pull-back of $D_z(h^t)=t^\beta x_1^{\alpha_1}\dots x_d^{\alpha_d}\varepsilon(\mathbf{x},t)$ where $\varepsilon(\mathbf{0},0)\neq 0$. Since $\mathbf{g}\in\mathbb{P}^{d+1}$ then $\tilde{\varepsilon}:=\varepsilon\circ\pi_{\mathbf{g}}\circ\pi^{q,w}_{t,j}(\widetilde{\mathbf{y}},t)$ also satisfies $\tilde{\varepsilon}(\mathbf{0},0)\neq 0$. The characteristic exponents do not depend on the particular $\mathbf{y}$ but only on J and the depth has decreased. Namely, by induction $Z^{\mathcal{W}}_{DL}((h^t_{q,j,\mathbf{g},w})_{(\mathbf{y},0)},(\tilde{\omega}_{q,j,\mathbf{g},w})_{(\mathbf{y},0)},T)$ satisfies the theorem. Observe also that the variety of such $\mathbf{y}$'s is $\mathbb{G}^{\#J}_{m,k}$.

Proposition 4.29. *The following equality holds:*

$$Z^{\mathcal{W},\mathbf{g},w}_{J,q,j}(T)=(\mathbb{L}-1)^{\#J}\mathbb{L}^{-(\nu_{\mathbf{g}}+A_{\mathbf{g}})}Z^{\mathcal{W}}_{DL}((h^t_{q,j,\mathbf{g},w})_{(\mathbf{y},0)},(\tilde{\omega}_{q,j,\mathbf{g},w})_{(\mathbf{y},0)},T),\tag{4.19}$$

where $_{(\mathbf{y},0)}$ *means the germ at this point.*

Proof. This result is a consequence of Proposition 4.35. □

Remark 4.30. We note that $(h^t_{q,j,\mathbf{g},w})_{(\mathbf{y},0)}$ is a $\mathcal{W}$-quasi-ordinary series in good coordinates but its depth is strictly less than $\operatorname{depth}(h^t_{q,j,\mathbf{g},w})$. Moreover its Newton polyhedron is obtained by projecting the Newton polyhedron of $h^t_{q,j,\mathbf{g},w}$ on the corresponding coordinate planes.

Step 10. Computations for h and $\mathbf{g}=\mu_1^{\mathbf{g}}\mathbf{w}_1^q+\dots+\mu_d^{\mathbf{g}}\mathbf{w}_d^q\in G^q$ such that $0<\mu_l^{\mathbf{g}}<1$ for some $l\in\{1,\dots,d\}$.

This case can be computed as in Step 9 since we may consider h as $\widetilde{\mathcal{W}}$-quasi-ordinary where $\widetilde{\mathcal{W}}=(1;0,\dots,0)$.

We are going to collect all the results in this section to compare motivic zeta functions for the functions h^t and h in order to obtain recurrence formulæ.

By the identities in (4.15),

$$
\begin{aligned}
(4.20)\quad & Z_{DL}^{\mathcal{W}}(h^t,\omega,T) = Z_A^{\mathcal{W}}(h^t,\omega,T)+\\
&+\sum_{q=1}^{r}\sum_{\substack{j=1\\ s_1^{j,q}=1}}^{u(q)}\Bigg(\sum_{\mathbf{g}\in G^{q,\alpha_{q,j}}}\mathbb{L}^{-(\nu_{\mathbf{g}}+A_{\mathbf{g}})}\Big(\sum_{w=1}^{w(q,j)}\big(Z_{DL}^{\mathcal{W}}(h^t_{q,j,\mathbf{g},w},\tilde{\omega}^t_{q,j,\mathbf{g},w},T)+\\
&+\sum_{\emptyset\subsetneq J\subset\{1,\ldots,d\}}(\mathbb{L}-1)^{\#J}Z_{DL}^{\mathcal{W}}((h^t_{q,j,\mathbf{g},w})_{(\mathbf{y},0)},(\tilde{\omega}_{q,j,\mathbf{g},w})_{(\mathbf{y},0)},T)\big)\Big)\Bigg),
\end{aligned}
$$

where $\nu_{\mathbf{g}}+A_{\mathbf{g}}=\frac{\alpha_{q,j}}{n_1^q}+\sum_{l=1}^{d}\mu_l^{\mathbf{g}}(p_l^q\nu_l+\bar{b}_l^q)$. In the same way the following identities hold

$$
\begin{aligned}
& Z_{DL}(h,\bar{\omega},T) = Z_{DL}^{A}(h,\bar{\omega},T)+\sum_{q=1}^{r}\sum_{\bar{j}=1}^{v(q)}(Z_{DL}(\bar{h}_{q,\bar{j}},\bar{\omega}_{q,\bar{j}},T)\\
(4.21)\quad &+\sum_{\mathbf{g}\in G^q,\mathbf{g}\neq\mathbf{g}^1}(\mathbb{L}^{-(\nu_{\mathbf{g}}+A_{\mathbf{g}})}(Z_{DL}^{\mathcal{W}}(\tilde{h}^t_{q,\mathbf{g},\bar{j}},\tilde{\omega}_{q,\mathbf{g},\bar{j}},T)+\\
&+\sum_{\emptyset\subsetneq J\subset\{1,\ldots,d\}}(\mathbb{L}-1)^{\#J}Z_{DL}^{\mathcal{W}}((\tilde{h}^t_{q,\mathbf{g},\bar{j}})_{(\mathbf{y},0)},(\tilde{\omega}_{q,\mathbf{g},\bar{j}})_{(\mathbf{y},0)},T)))).
\end{aligned}
$$

where $v(q)=\sum_{j=1}^{u(q)}w(q,j)$ and $\nu_{\mathbf{g}}+A_{\mathbf{g}}=\sum_{l=1}^{d}\mu_l^{\mathbf{g}}(p_l^q\nu_l+\bar{b}_l^q)$.

PROOF OF THEOREM 4.3. The starting point of the induction has been done in Step 1.

Let us attack the general case. Since h^t and h (as K-quasi-ordinary power series) have the same characteristic exponents then by Lemma 3.34 there exist one-to-one correspondences between the sets of faces of their Newton polyhedra and their sets of roots on such faces for h^t and h. Assume that under these bijections the root $\boldsymbol{\beta}^q_{j,w}(t)$, where $s_1^{q,j}=1$, corresponds to $\bar{\beta}^q_{\bar{j}}$, namely the polynomial $z^{n_1^q}-\boldsymbol{\beta}^q_{j,w}(t)x_1^{b_1^q}\ldots x_d^{b_d^q}$ corresponds to $\bar{z}^{n_1^q}-\bar{\beta}^q_{\bar{j}}\bar{x}_1^{b_1^q}\ldots\bar{x}_d^{b_d^q}$. Let $h^t_{\boldsymbol{\beta}}$ and $\bar{h}_{q,\bar{j}}$ denote the pull-back of h^t and h under the corresponding Newton maps associated with the roots $\boldsymbol{\beta}^q_{j,w}(t)$ and $\bar{\beta}^q_{\bar{j}}$.

(1) By Lemma 3.34, the characteristic exponents of $h^t_{\boldsymbol{\beta}}$ and $\bar{h}_{q,\bar{j}}$ are equal.
(2) For each $\mathbf{g}\in G^q$ (resp. $\mathbf{g}\in G^{q,\alpha_{q,j}}$) the K-quasi-ordinary power series $h\circ\pi_{\mathbf{g}}$ (resp. $h^t\circ\pi_{\mathbf{g}}$) has the same characteristic exponents as h (resp. h^t). This is Remark 3.35.
(3) Applying (1) and (2) for each $\mathbf{g}\in G^q\setminus\{\mathbf{g}^1\}$, then $\tilde{h}_{q,\mathbf{g},\bar{j}}$ has the same characteristic exponents as $\bar{h}_{q,\bar{j}}$ (recall that $\bar{h}_{q,\bar{j}}$ is obtained in Step 8).
(4) In the same way, for each $\mathbf{g}\in G^{q,\alpha_{q,j}}$, $h^t_{q,j,\mathbf{g},w}$ has the same characteristic exponents as $h^t_{\boldsymbol{\beta}}$ which has the same as $\bar{h}_{q,\bar{j}}$. In all these cases, the depths of all these quasi-ordinary power series are smaller than depth(h). In particular we can follow by induction.

(5) For the statement of the Theorem the terms $\mathbb{L}^{-\nu}$ are not essential then we have that

$$Z_{DL}^{\mathcal{W}}(\tilde{h}^t_{q,\mathbf{g},\bar{j}}, \tilde{\omega}_{q,\mathbf{g},\bar{j}}, T) \in \mathbb{Z}[\mathbb{L}, \mathbb{L}^{-1}, (1-\mathbb{L}^{-\nu}T^N)^{-1}][T],$$

where $(N, \nu) \in CP(\bar{h}_{q,\bar{j}}, \bar{\omega}_{q,\bar{j}})$.

□

PROOF OF THEOREM 4.9. The starting point of the induction was done in Step 1.

For the general case we consider the formulæ (4.20) and (4.21). In this case the restriction $s_1^{j,q} = 1$ is empty and the series appearing in the second line of (4.20) are $\widetilde{\mathcal{W}}'$-quasi-ordinary for some weights and they have depth less than h. The series in the third line are only $\mathcal{W}$-quasi-ordinary.

Because of Theorem 4.3 we can compute $\chi_{\mathrm{top}}(\bullet, (\mathbb{L}^{-s}))$ for each term. For the second line terms we deduce by induction that

$$\chi_{\mathrm{top}}(\mathbb{L}^{-(\nu_{\mathbf{g}}+A_{\mathbf{g}})} Z_{DL}^{\mathcal{W}}(h^t_{q,j,\mathbf{g},w}, \tilde{\omega}_{q,j,\mathbf{g},w}, (\mathbb{L}^{-s}))) = Z_{\mathrm{top},0}(\bar{h}_{q,\bar{j}}, \omega_{q,\bar{j}}, s).$$

By Theorem 4.3 and (4.19), the terms in the third line vanish when applying $\chi_{\mathrm{top}}(\bullet(\mathbb{L}^{-s}))$. □

3. Zeta functions along strata

To finish the proof of Theorems 4.3 and 4.9 we must prove the identity (4.19) in Proposition 4.29. We will proceed by induction in *families* of $\mathcal{W}$-quasi-ordinary power series.

DEFINITION 4.31. Let $h^t \in k[[t]][[\mathbf{x}, \mathbf{y}]][z]$ be a $\mathcal{W}$-quasi-ordinary power series, $\mathbf{x} = (x_1, \ldots, x_d)$, $\mathbf{y} = (y_1, \ldots, y_e)$ which is bounded for the $\mathbf{y}$-variables. We say that h^t is *a family of $\mathcal{W}$-quasi-ordinary power series* if for any $\mathbf{y}^0 \in \mathbf{G}^e_{k,m}$, the well-defined germ $h^t_{\mathbf{y}^0} := h^t(\mathbf{x}, \mathbf{y} + \mathbf{y}^0)$ is also $\mathcal{W}$-quasi-ordinary power series with the same characteristic monomials.

DEFINITION 4.32. The Denef-Loeser zeta function of a family of $\mathcal{W}$-quasi-ordinary power series and a form ω satisfying the support condition (1.3) is defined by

$$Z_{DL}^{\mathcal{W},\mathbf{y}}(h^t, \omega, T) := T^{\theta} \sum_{n\in\mathbb{N}} \left(\sum_{m\in\mathbb{N}} \mathbb{L}^{-m} \mu_X(V^t_{n,m}) \right) T^n \in \widehat{\mathcal{M}}_k[[T]],$$

$$Z_{\mathrm{top},0}^{\mathcal{W}}(h^t, \omega, s) := \chi_{\mathrm{top}}(Z_{DL}^{\mathcal{W}}(h^t, \omega, \mathbb{L}^{-s})).$$

where we are considering arcs having their origin along $\{0\} \times \mathbf{G}^e_{k,m}$.

REMARK 4.33. The powers of the variables in $\mathbf{y}$ for ω are negligible.

LEMMA 4.34. *$Z_{DL}^{\mathcal{W},\mathbf{y}}(h^t, \omega, T) = Z_{DL}^{\mathcal{W},\mathbf{y}}(\tilde{h}^t, \omega, T)$ where $\tilde{h}^t(\mathbf{x}, \mathbf{y}) = h^t(\mathbf{x}, \mathbf{y}^{\mathbf{n}})$.*

PROPOSITION 4.35. *Let h^t be a family of $\mathcal{W}$-quasi-ordinary power series and ω a form satisfying the support condition (1.3). Then for all $\mathbf{y}^0 \in \mathbf{G}^e_{k,m}$,*

$$Z_{DL}^{\mathcal{W},\mathbf{y}}(h^t, \omega, T) = (\mathbb{L}-1)^e Z_{DL}^{\mathcal{W}}(h^t_{\mathbf{y}^0}, \omega, T).$$

PROOF. The proof is by induction on the depth of h^t; if depth $h^t = 0$, the result is trivial. We compute $Z_{DL}^{\mathcal{W},\mathbf{y}}(h^t, \omega, T)$ following the Newton polyhedron and the Newton maps.

We begin with the A-part. If we consider a vertex, it is related to a monomial $t^\alpha \mathbf{x}^\mathbf{n} \mathbf{y}^\mathbf{m} z^n$. It is clear that both terms in the equality differ by an $(\mathbb{L}-1)^e$ factor. Let us fix an edge of the Newton polytope and a factor of the principal part which will give the key for the Newton map, i.e., we fix some factor $(z^n - \beta t^\alpha \mathbf{x}^\mathbf{n} \mathbf{y}^\mathbf{m})^m$.

The mapping

$$(\mathbf{x}, \mathbf{y}, z) \mapsto (\mathbf{x}, y_1^n, \dots, y_e^n, y_1^{m_1} \dots y_e^{m_e} z)$$

does not change $Z_{DL}^{\mathcal{W},\mathbf{y}}(h^t, \omega, T)$. Thus we may suppose that the factor is $(z^n - \beta t^\alpha \mathbf{x}^\mathbf{n})^m$. Since the computations for the A-part depend only on this Newton principal part, it is clear that both terms in the equality differ by an $(\mathbb{L}-1)^e$ factor.

Let us consider now the B-part. For the sake of simplicity we can suppose that all coordinates of $\mathbf{n}$ are non-zero. Following the argument of Step 9, the zeta function corresponding to the B-part is the zeta function associated with a power series $\tilde{h}^t$ and a form $\tilde{\omega}$ where we must consider arcs based at $\mathbb{A}_k^d \times \mathbf{G}_{k,m}^e$. We can perform a stratification of this space and decompose the result in a sum of zeta functions of families of $\mathcal{W}$-quasi-ordinary power series, where variables corresponding to $J \subset \{1, \dots, d\}$, pass to $\mathbf{y}$. Since these power series have strictly smaller depth we also obtain the $(\mathbb{L}-1)^e$ factor. □

The Theorems 4.3 and 4.9 are proved.

EXAMPLE 4.36. Consider the quasi-ordinary power series $h := (z^2 - x_1^3 x_2^5)^2 - 4x_1^7 x_2^{13} \in \mathbb{C}[[x_1, x_2]][z]$ and the differential form $\bar{\omega} =: dx_1 \wedge dx_2 \wedge dz$, the pair $(h, \bar{\omega})$ satisfies the support condition. Its depth equals 2, it has only one compact 1-dimensional face γ_1 $(q = 1)$ with only one Newton component $(v(q) = v(1) = 1)$. For the face, in the dual the linearly independent vectors are $\mathbf{w}_1 := (2, 0, 3)$ and $\mathbf{w}_2 := (0, 2, 5)$. Thus the set G^1 has two elements $\mathbf{g}^1 := (2, 2, 8)$ and $\mathbf{g} := (1, 1, 4)$. Applying (4.15),

$$Z_{DL}(h, \bar{\omega}, T) = Z_{DL}^A(h, \bar{\omega}, T) + Z^{\bar{\gamma}_1, \mathbf{g}^1, 1}(T) + Z^{\bar{\gamma}_1, \mathbf{g}, 1}(T),$$

where $Z_{DL}^A(h, \bar{\omega}, T)$ has been explicitly described in (4.14) and Remark 3.18.

In **Step 8** we have described how to compute $Z^{\bar{\gamma}_1, \mathbf{g}^1, 1}(T)$, see (4.16). The Newton map $\pi_{1,1}$ is defined by $x_1 = y_1^2$, $x_2 = y_2^2$ and $z = (z_1 + 1) y_1^3 y_2^5$. Then the pull-back of h is given by $\bar{h}_{1,1} := y_1^{12} y_2^{20}(((z_1+1)^2 - 1)^2 - 4y_1^2 y_2^6)$ which is quasi-ordinary with depth equals 1, (i.e. $\bar{h}_{1,1}$ is non-degenerate). The pull-back $\bar{\omega}_{1,1}$ of the differential form is $y_1^4 y_2^6 dy_1 \wedge dy_2 \wedge dz_1$, up to a constant factor. Then $Z^{\bar{\gamma}_1, \mathbf{g}^1, 1}(T) = Z_{DL}(\bar{h}_{1,1}, \bar{\omega}_{1,1}, T)$.

To compute $Z^{\bar{\gamma}_1, \mathbf{g}, 1}(T)$ we follow **Step 10**, see also **Step 9**. Under this process we first consider the $\mathbb{C}[t]$-morphism $\pi_\mathbf{g}$ given by $x_1 = t\bar{y}_1$, $x_2 = t\bar{y}_2$ and $z = t^4 \bar{z}_1$, (see Lemma 4.25) and after the map $\pi_{1,t}^1$ defined by $\bar{y}_1 = y_1^2$, $\bar{y}_2 = y_2^2$ and $\bar{z} = (z_1 + 1) y_1^3 y_2^5$. The pull-back $\bar{h}_2^t := h \circ \pi_\mathbf{g} \circ \pi_{1,t}^1$ of h under the composition map is $\bar{h}_2^t = t^{16} y_1^{12} y_2^{20}(((z_1+1)^2 - 1)^2 - 4t^4 y_1^2 y_2^6)$. Then $\bar{h}_2^t$ has depth equals to 1. Applying

(4.20) the following identity holds

$$Z^{\bar{\gamma}_1,\mathbf{g},1}(T) = \mathbb{L}^{-6}\left(Z_{DL}^{\mathcal{W}}(\bar{h}_2^t,\tilde{\omega}_{\mathbf{g}},T) + \sum_{\emptyset\subsetneq J\subset\{1,2\}}(\mathbb{L}-1)^{\#J}Z_{DL}^{\mathcal{W}}((\bar{h}_2^t)_{(\mathbf{y},0)},(\tilde{\omega}_{\mathbf{g}})_{(\mathbf{y},0)},T)\right),$$

where $6 = \nu_{\mathbf{g}} + A_{\mathbf{g}}$ and $\tilde{\omega}_{\mathbf{g}}$ in this case coincides with $\bar{\omega}_{1,1}$. Thus we can apply induction on the depth.

CHAPTER 5

Consequences of the main theorems

In this section we describe several consequences of the previous results. Firstly the local Denef-Loeser motivic zeta function of a quasi-ordinary power series $h \in k[[\mathbf{x}]][z]$ essentially depends on its essential variables. Secondly, in the $\mathbf{ev}(h) = 1$ case, there exists a recursive formula to compute $Z_{DL}(h, \omega, T)$ for any differential form such that (h, ω) satisfies the support condition 1.3. If $\mathbf{ev}(h) > 1$ the recursive formula can be only obtained at the level of the topological zeta function. Another, even more interesting, consequence is the inductive definition of a set of *strong candidate poles* $SCP(h, \omega)$ which is in general smaller than the set $CP(h, \omega)$ of candidate poles. Moreover the local Denef-Loeser motivic zeta function satisfies

$$Z_{DL}(h, \omega, T) \in \mathbb{Z}[\mathbb{L}, \mathbb{L}^{-1}, (1 - \mathbb{L}^{-\nu}T^N)^{-1}][T]_{(N,\nu)\in SCP(h,\omega)}.$$

Furthermore the set $SCP(h, \omega)$ is a subset of the union of all strong candidate poles of the of all the transversal sections of (h, ω). We will use this fact to prove the monodromy conjecture by induction on $\mathbf{ev}(h)$.

1. Essential variables

Let $h \in k[[\mathbf{x}]][z]$ be a quasi-ordinary power series. For any subset $\emptyset \subsetneq I \subsetneq \{1, \ldots, d\}$, let k_I be an algebraic closure of the quotient field of the domain $k[[x_i]]_{i\in I}$. Thus $h_I \in k_I[[\widehat{\mathbf{x}}_I]][z]$ is a k_I-quasi-ordinary power series, where $\widehat{\mathbf{x}}_I$ are the variables not in I.

Assume $h \in k[[\mathbf{x}]][z]$ is a quasi-ordinary power series with $\mathbf{ev}(h) = e < d$. Relabeling the variables we assume that $x_1, \ldots, x_e$ are the essential variables. Set $I = \{e+1, \ldots, d\}$. Let $\omega = \prod_{l=1}^{d} x_j^{\nu_j - 1} d\mathbf{x} \wedge dz$ be a differential form such that (h, ω) satisfies the support condition 1.3. Whenever the variable x_j is not essential then the support condition implies $\nu_j = 1$. In particular if $\omega_I = \prod_{l=1}^{e} x_j^{\nu_j - 1} d\widehat{\mathbf{x}}_I \wedge dz$ then the pair (h_I, w_I) also satisfies the support condition 1.3 over the field k_I.

Corollary 5.1. *Given the previous conditions, the following formal identity holds:*

$$Z_{DL}(h, \omega, T) = \mathbb{L}^{-(d+1-e)} Z_{DL}(h_I, w_I, T).$$

Here formal means that in the RHS, resp. LHS, the varieties are over k, resp. k_I. The proof of the corollary follows essentially from the proof of theorems 4.3 and 4.9. In fact Newton maps do affect the non-essential variables.

2. Curve case

Suppose a $\mathcal{W}$-quasi-ordinary power series h^t has $\mathbf{ev}(h^t) = 1$. After relabeling the variables x_i, assume $D_z(h^t) = t^{\alpha} x_1^{\alpha_1} \varepsilon(t, \mathbf{x})$, with $\varepsilon(0, \mathbf{0}) \neq 0$. We assume the same for the quasi-ordinary power series h. Then the compact faces of their Newton polyhedra are contained in 2-dim plane $x_2 = \ldots = x_d = 0$. In particular, in the dual space one has the same property. The dual hyperplanes $l^t_{q,j}$ are defined by $\alpha_{q,j} + s_1^{q,j}(b_1^q v_1 - n_1^q v_{d+1}) = 0$, where $\gcd(b_1^q, n_1^q) = 1$. The dual hyperplanes are determined by the linearly independent vectors $\mathbf{w}_1^q$ and $\mathbf{w}_l^q = e_l$ for $l = 2, \ldots, d$. Therefore the sets G^q and $G^{q,\alpha_{q,j}}$ have one element. Then we can conclude that

(5.1)

$$Z_{DL}^{\mathcal{W}}(h^t, \omega, T) = Z_A^{\mathcal{W}}(h^t, \omega, T) + \sum_{q=1}^{r} \sum_{\substack{j=1 \\ s_1^{q,j}=1}}^{u(q)} \mathbb{L}^{-(\nu_{\mathbf{g}} + A_{\mathbf{g}})} \sum_{w=1}^{w(q,j)} Z_{DL}^{\mathcal{W}}(h^t_{q,j,\mathbf{g},w}, \tilde{\omega}^t_{q,j,\mathbf{g},w}, T),$$

$$Z_{DL}(h, \bar{\omega}, T) = Z_{DL}^A(h, \bar{\omega}, T) + \sum_{q=1}^{r} \sum_{\bar{j}=1}^{v(q)} Z_{DL}(\bar{h}_{q,\bar{j}}, \omega_{q,\bar{j}}, T),$$

where $\nu_{\mathbf{g}} + A_{\mathbf{g}} = \frac{\alpha_{q,j}}{n_1^q} + \mu_1^{\mathbf{g}}(p_1^q \nu_l + \bar{b}_1^q)$ and $v(q) = \sum_{j=1}^{u(q)} w(q,j)$.

COROLLARY 5.2. *For a quasi-ordinary power series h with $\mathbf{ev}(h) = 1$ and a form ω such that (h, ω) satisfies condition (1.3) the second line in equations (5.1) can be used as recursive formula to compute $Z_{DL}(h, \bar{\omega}, T)$.*

A general formula for the topological zeta function will be given in Theorem 5.3.

3. The topological zeta function

Assume the quasi-ordinary power series $h(\mathbf{x}, z) := \prod_{l=1}^d x_l^{N_l} f(\mathbf{x}, z) u(\mathbf{x}, z)$, $N_l \in \mathbb{N}$, is given in a good system of coordinates and that the differential form $\omega = \prod_{j=1}^d x_j^{\nu_j - 1} dx_1 \wedge \ldots \wedge dx_d \wedge dz$, $\nu_j \geq 1$, satisfies condition (1.3). Assume that $\gamma_1, \ldots, \gamma_r$ (resp. $\tau_0, \ldots, \tau_r$) are the 1-dimensional (resp. 0-dimensional) compact faces of its Newton polyhedron. Assume for each $q = 1, \ldots, r$, on γ_q there are exactly $v(q)$ distinct roots. Let $\pi_{q,j}$ be the Newton map associated with the Newton component f_j^q of f.

THEOREM 5.3. *Under the above conditions the following equality holds*

$$Z_{\text{top},0}(h, w, s) = \sum_{l=0}^{r} J_{\tau_l}(h, \omega, s) - \sum_{q=1}^{r} v(q) J_{\gamma_q}(h, \omega, s)$$
$$+ \sum_{q=1}^{r} \text{mult}(\Delta_{\gamma_q}) \sum_{j=1}^{v(q)} Z_{\text{top},0}(h \circ \pi_{q,j}, w \circ \pi_{q,j}, s).$$

Moreover if h has a non-degenerate principal part then for each $q = 1, \ldots, r$ the integer number $v(q)$ is $V(\gamma_q)$ and for any $j = 1, \ldots, v(q)$, the following identity holds

$$Z_{\text{top},0}(h \circ \pi_{q,j}, w \circ \pi_{q,j}, s) = \frac{J_{\gamma_q}(h, \omega, s)}{(s+1)\,\text{mult}(\Delta_{\gamma_q})}.$$

PROOF. The same ideas as in the construction of the above section give $Z_{DL}(h,\omega,T)$ as

$$Z_{DL}(h,\omega,T) = Z_{DL}^{A}(h,\omega,T) + \sum_{q=1}^{r}\sum_{j=1}^{v(q)}\sum_{\mathbf{g}\in G^q} Z^{\gamma_q,\mathbf{g},j}(T),$$

see equation (4.15). The contribution of the A-part was given in equation (3.16).

CONTRIBUTION OF THE B-PART. Fix a 1-dim face γ_q, a root β_j^q of γ_q and $\mathbf{g}$ in G^q.

(1) If $\mathbf{g}=\mathbf{g}^1$ then $Z^{\gamma_q,\mathbf{g}^1,j}(T) = Z_{DL}(\tilde{h},\tilde{\omega},T)$, where $\tilde{h} = h\circ\pi_{q,j}$ and $\tilde{\omega} = \omega\circ\pi_{q,j}$, see equation (4.16) in Step 8.
(2) If $\mathbf{g}\neq\mathbf{g}^1$ then $Z^{\gamma_q,\mathbf{g},j}(T) = Z_1^{\gamma_q,\mathbf{g},j}(T) + \sum_{\emptyset\subsetneq J\subset\{1,\ldots,d\}} Z_J^{\gamma_q,\mathbf{g},j}(T)$, see Step 9.

Recall that the following properties hold.

(1) $\chi_{\text{top}}(Z_1^{\gamma_q,\mathbf{g},j}(\mathbb{L}^{-s})) = Z_{\text{top},0}(h\circ\pi_{q,j},\omega\circ\pi_{q,j},s)$, see equation (4.18).
(2) $\chi_{\text{top}}(Z_J^{\gamma_q,\mathbf{g},j}(\mathbb{L}^{-s})) = 0$, see equation (4.19).

Hence the contribution to $Z_{\text{top},0}(h,w,s)$ of the B-part is

$$\sum_{q=1}^{r}\sum_{j=1}^{v(q)}\sum_{\mathbf{g}\in G^q} Z_{\text{top},0}(h\circ\pi_{q,j},\omega\circ\pi_{q,j},s).$$

Finally the cardinality of G^q is $\text{mult}(\Delta_{\gamma_q})$.

The second part of the theorem follows from the results of Chapter 2, in particular from Lemma 2.5. □

ALGORITHM. Our method gives an effective algorithm to compute $Z_{\text{top},0}(h,w,s)$. After a Newton map $\pi_{q,j}$, $h\circ\pi_{q,j}(\mathbf{y},z_1) = y_1^{N_1}\cdots y_d^{N_d}\tilde{f}(\mathbf{y},z_1)\tilde{u}(\mathbf{y},z_1)$ where $\tilde{f}\in k[[\mathbf{y}]][z_1]$ is a z_1-quasi-ordinary power series. Hence $\tilde{f}(\mathbf{0})=0$ and $N_l = \inf_{\mathbf{x}\in\Gamma(h)}\{\mathbf{w}_l\cdot\mathbf{x}\}$ for any $l=1,\ldots,d$. Apply again Theorem 5.3 until depth 0. This algorithm can be implemented in Maple based on a Maple program made by K. Hoornaert and D. Loots for the non-degenerate case, [**29**].

4. A special candidate pole

In this paragraph we will show that some candidate poles disappear when we add their local contributions. For a candidate pole $s_0 = (N,\nu)$ coming from a compact 1-dimensional face γ, its *local contribution* is defined as the sum of the A-part of the local Denef-Loeser motivic zeta function which corresponds to γ and its two vertices plus the sum of the contributions of the B-part. The latter is the contribution of the highest vertex in the new Newton polyhedron after all Newton maps of Newton components associated with γ and at all possible new quasi-ordinary power series.

Consider the following formal power series

$$\begin{aligned} h^t(\mathbf{x},z) &:= x_1^{N_1}\ldots x_{d-1}^{N_{d-1}} z^{n-n_1 m}(z^{n_1} - \beta t^{\alpha} x_1^{b_1}\ldots x_{d-1}^{b_{d-1}} x_d)^m + \ldots, \\ h(\mathbf{x},z) &:= x_1^{N_1}\ldots x_{d-1}^{N_{d-1}} z^{n-n_1 m}(z^{n_1} - \beta x_1^{b_1}\ldots x_{d-1}^{b_{d-1}} x_d)^m + \ldots, \end{aligned}$$

where the other terms appears *behind* their Newton polyhedra and $\alpha \in \mathbb{P}, \beta \in \mathbf{G}_{m,k}$. Assume $h^t \in k[[t]][[\mathbf{x}]][z]$ is $\mathcal{W}$-quasi-ordinary and $h \in k[[\mathbf{x}]][z]$ is a quasi-ordinary power series.

Consider the differential form $\omega := x_1^{\nu_1-1} \dots x_{d-1}^{\nu_{d-1}-1} dx_1 \wedge \dots \wedge dx_d \wedge dz$, in such a way (h^t, ω) and (h, ω) satisfy the support condition 1.3. In other words the pair $(N_d, b_d) = (0, 1)$. In this section the local contribution of the candidate pole $s_0 := 1 - \mathbb{L}^{-(n_1+1)}T^n$ which corresponds with the d-th coordinate of the first characteristic exponent of h^t and h is computed. Under the above conditions we have the following result.

PROPOSITION 5.4. *If there is no pair (N_i, b_i) equal to $(0, 1)$ other than (N_d, b_d) then $s_0 = 1 - \mathbb{L}^{-(n_1+1)}T^n$ does not appear in the denominator of the local contribution of s_0 to $Z_{DL}^{\mathcal{W}}(h^t, \omega, T)$ and $Z_{DL}(h, \omega, T)$.*

PROOF. The compact faces of $\Gamma(h)$ and $\Gamma^t(h^t)$ are the same and consist of a face γ with vertices $\tau_0 := (N_1 + mb_1, \dots, N_{d-1} + mb_{d-1}, m, n - n_1 m)$ and $\tau_1 := (N_1, \dots, N_{d-1}, 0, n)$.

Let $\eta := b_1 v_1 + \dots b_{d-1} v_{d-1} + v_d - n_1 v_{d+1}$. In the dual space, the dual decomposition induced by $\Gamma(h)$ has only three convex rational cones $\Delta_{\tau_1} = \{\eta > 0\}$, $\Delta_{\tau_0} = \{\eta < 0\}$ and $\Delta_\gamma = \{\eta = 0\}$. In the same way, the decomposition induced by $\Gamma^t(h)$ is given by the convex rational polyhedron $\Delta^t_{\tau_1} = \{\eta + \alpha > 0\}$, and the cones $\Delta^t_{\tau_0} = \{\eta + \alpha < 0\}$ and $\Delta^t_\gamma = \{\eta + \alpha = 0\}$. Since there is only one exponent α of t, convex polyhedra delimited by parallel hyperplanes do not exist.

To compute the contributions of the candidate pole $s_0 := 1 - \mathbb{L}^{-(n_1+1)}T^n$ we multiply $Z_{DL}^{\mathcal{W}}(h^t, \omega, T)$ and $Z_{DL}(h, \omega, T)$ by s_0 and simplify the result under the condition $\mathbb{L}^{-(n_1+1)}T^n = 1$.

The contribution coming from the Newton polyhedra of h and h^t written in terms of generating functions are:

$$
\begin{aligned}
Z_{DL}^{\Delta^t_{\tau_0}, \mathcal{W}}(T) &= T^{m\alpha}(\mathbb{L}-1)^{d+1}\mathbb{L}^{-(d+1)}\Phi_{\Delta^t_{\tau_0}}(\mathbf{y}),\\
Z_{DL}^{\Delta^t_{\tau_1}, \mathcal{W}}(T) &= (\mathbb{L}-1)^{d+1}\mathbb{L}^{-(d+1)}\Phi_{\Delta^t_{\tau_0}}(\mathbf{a}),\\
Z_{DL}^{\Delta^t_{\gamma}, \mathcal{W}}(T) &= \mathbb{L}^{-(d+1)}\left((\mathbb{L}-1)^{d+1} - (\mathbb{L}-1)^d\right)\Phi_{\Delta^t_\gamma}(\mathbf{a}),\\
Z_{DL}^{\Delta_{\tau_0}}(T) &= (\mathbb{L}-1)^{d+1}\mathbb{L}^{-(d+1)}\Phi_{\Delta_{\tau_0}}(\mathbf{y}),\\
Z_{DL}^{\Delta_{\tau_1}}(T) &= (\mathbb{L}-1)^{d+1}\mathbb{L}^{-(d+1)}\Phi_{\Delta_{\tau_0}}(\mathbf{a}),\\
Z_{DL}^{\Delta_{\gamma}}(T) &= \mathbb{L}^{-(d+1)}\left((\mathbb{L}-1)^{d+1} - (\mathbb{L}-1)^d\right)\Phi_{\Delta_\gamma}(\mathbf{a}),
\end{aligned}
$$

where $\mathbf{a} := (a_1, \dots, a_{d-1}, a_d, a_{d+1}) = (\mathbb{L}^{-\nu_1}T^{N_1}, \dots, \mathbb{L}^{-\nu_{d-1}}T^{N_{d-1}}, \mathbb{L}^{-1}, \mathbb{L}^{-1}T^n)$ and $\mathbf{y} := (y_1, \dots, y_d, y_{d+1})$ is equal to

$$(\mathbb{L}^{-\nu_1}T^{N_1+mb_1}, \dots, \mathbb{L}^{-\nu_{d-1}}T^{N_{d-1}+mb_{d-1}}, \mathbb{L}^{-1}T^m, \mathbb{L}^{-1}T^{n-n_1 m}).$$

If $\Delta_{\mathbf{1}}$ denotes the positive cone $\mathbb{P}^{d+1}$ then the following identities among indicator functions hold: $[\Delta^t_{\tau_1}] = [\Delta_{\mathbf{1}}] - [\Delta^t_{\tau_0}] - [\Delta^t_\gamma]$ and $[\Delta_{\tau_1}] = [\Delta_{\mathbf{1}}] - [\Delta_{\tau_0}] - [\Delta_\gamma]$. Summing and looking only at the terms where $1 - \mathbb{L}^{-(n_1+1)}T^n$ appears, the contribution of s_0 is given by

$$\text{(5.2)} \qquad \mathbb{L}^{-(d+1)}(\mathbb{L}-1)^{d+1}\left(-A^t(\mathbf{a}) + T^{m\alpha}A^t(\mathbf{y}) - \frac{\mathbb{L}^{-1}\Phi_{\Delta^t_\gamma}(\mathbf{a})}{(1-\mathbb{L}^{-1})}\right), \text{ for } h^t \text{ and}$$

$$\mathbb{L}^{-(d+1)}(\mathbb{L}-1)^{d+1}\left(-A(\mathbf{a})+A(\mathbf{y})-\frac{\mathbb{L}^{-1}\Phi_{\Delta_\gamma}(\mathbf{a})}{(1-\mathbb{L}^{-1})}\right),\quad \text{for } h,$$

where

$$A^t(\mathbf{x}):=\frac{\sum_{\boldsymbol{\beta}\in G_0^t}\mathbf{x}^{\boldsymbol{\beta}}}{(1-x_{d+1})\prod_{l=1}^{d}(1-\mathbf{x}^{\mathbf{w}_l})},\qquad A(\mathbf{x}):=\frac{\sum_{\boldsymbol{\beta}\in G_0}\mathbf{x}^{\boldsymbol{\beta}}}{(1-x_{d+1})\prod_{l=1}^{d}(1-\mathbf{x}^{\mathbf{w}_l})},$$

and G_0^t (resp. G_0) is the fundamental set of $\Delta_{\tau_0}^t$ (resp. Δ_{τ_0}).

We will compute the terms

$$A_1^t:=(1-\mathbb{L}^{-(n_1+1)}T^n)\left(-A^t(\mathbf{a})+T^{m\alpha}A^t(\mathbf{y})-\frac{\mathbb{L}^{-1}\Phi_{\Delta_\gamma^t}(\mathbf{a})}{(1-\mathbb{L}^{-1})}\right)$$

and $A_1:=(1-\mathbb{L}^{-(n_1+1)}T^n)\left(-A(\mathbf{a})+A(\mathbf{y})-\frac{\mathbb{L}^{-1}\Phi_{\Delta_\gamma}(\mathbf{a})}{(1-\mathbb{L}^{-1})}\right)$ under the given condition $\mathbb{L}^{-(n_1+1)}T^n=1$.

Consider the vertex τ_1. The simplification $\mathbb{L}^{-(n_1+1)}T^n=1$ gives $a_d^{n_1}a_{d+1}=1$. Define $w_j:=a_j^{p_j}a_{d+1}^{\bar{b}_j}$, $1\le j\le d-1$, and $c_i:=\gcd(n_1,b_i)$ and $p_i:=\frac{n_1}{c_i}$, $\bar{b}_i:=\frac{b_i}{c_i}$, $i=1,\dots,d-1$. Thus $w_j=\mathbb{L}^{-(\nu_jp_j+\bar{b}_j)}T^{p_jN_j+n\bar{b}_j}=\mathbb{L}^{(b_j-\nu_j)p_j}T^{p_jN_j}$, and $w_j=u_j^{p_j}$ with $u_j:=\mathbb{L}^{b_j-\nu_j}T^{N_j}$.

The elements of Δ_{τ_0} are written as

$$\sum_{j=1}^{d-1}\mu_j(p_je_j+\bar{b}_je_{d+1})+\mu_d(n_1e_d+e_{d+1})+\mu_{d+1}e_{d+1}\in\mathbb{P}^{d+1},\mu_i\in\mathbb{Q}_{>0}.$$

The elements of $\Delta_{\tau_0}^t$ can be written as

$$\sum_{j=1}^{d-1}\mu_j(p_je_j+\bar{b}_je_{d+1})+\mu_d(n_1e_d+e_{d+1})+\mu_{d+1}e_{d+1}\in\mathbb{P}^{d+1},\mu_i\in\mathbb{Q}_{>0},\mu_{d+1}>\frac{\alpha}{n_1}.$$

To parametrize the sets G_0^t and G_0 we may replace μ_d by μ_d+s, $s\in\mathbb{Z}$, because of the substitution $a_d^{-n_1}=a_{d+1}$. Thus the elements in G_0^t (and in G_0 when $\alpha=0$) are obtained with

$$(\mu_1,\dots,\mu_{d-1},\mu_d,\mu_{d+1})=\left(\frac{i_1}{p_1},\dots,\frac{i_{d-1}}{p_{d-1}},-\frac{\alpha+k+\sum_{j=1}^{d-1}i_jb_j}{n_1},\frac{\alpha+k}{n_1}\right),$$

$i_j=1,\dots,p_j$, $j=1,\dots,d-1$, and $k=1,\dots,n_1$. Therefore $A^t(\mathbf{a})$ (resp. $A(\mathbf{a})$) multiplied by $(1-a_d^{n_1}a_{d+1})=1-\mathbb{L}^{-(n_1+1)}T^n$ is:

$$\frac{\sum_{i_1=1}^{p_1}\cdots\sum_{i_{d-1}=1}^{p_{d-1}}\sum_{k=1}^{n_1}a_d^{-(\alpha+k+\sum_{j=1}^{d-1}i_jb_j)}\prod_{j=1}^{d-1}a_j^{i_j}}{(1-a_{d+1})\prod_{j=1}^{d-1}(1-a_j^{p_j}a_{d+1}^{\bar{b}_j})}=\frac{a_d^{-\alpha}(\sum_{k=1}^{n_1}a_d^{-k})\prod_{j=1}^{d-1}\sum_{i_j=1}^{p_j}(a_ja_d^{-b_j})^{i_j}}{(1-a_{d+1})\prod_{j=1}^{d-1}(1-a_j^{p_j}a_{d+1}^{\bar{b}_j})}=$$

$$= -\frac{a_d^{-\alpha}(1-a_d^{-n_1})\prod_{j=1}^{d-1}\left(a_ja_d^{-b_j}(1-(a_ja_d^{-b_j})^{p_j})\right)}{(1-a_d)(1-a_{d+1})\prod_{j=1}^{d-1}\left((1-a_ja_d^{-b_j})(1-a_j^{p_j}a_{d+1}^{\bar{b}_j})\right)}. \tag{5.3}$$

The first summand $A^t(\mathbf{a})(1-a_d^{n_1}a_{d+1})$ (resp. $A(\mathbf{a})(1-a_d^{n_1}a_{d+1})$ when $\alpha=0$) is:

$$-\frac{a_d^{-\alpha}\prod_{j=1}^{d-1}u_j}{(1-a_d)\prod_{j=1}^{d-1}(1-u_j)}. \tag{5.4}$$

Consider the term from γ. The elements of Δ_γ^t (resp. Δ_γ when $\alpha=0$) are

$$\sum_{j=1}^{d-1}\mu_j'(p_je_j+\bar{b}_je_{d+1})+\ldots+\mu_d'(n_1e_d+e_{d+1})+\frac{\alpha}{n_1}e_{d+1}\in\mathbb{P}^{d+1},\mu_i'\in\mathbb{Q}_{>0}.$$

After simplifying, the fundamental set G_γ^t (resp. G_γ) of Δ_γ^t (resp. Δ_γ) is obtained from

$$(\mu_1',\ldots,\mu_{d-1}',\mu_d')=\left(\frac{i_1}{p_1},\ldots,\frac{i_{d-1}}{p_{d-1}},-\frac{\alpha+\sum_{j=1}^{d-1}i_jb_j}{n_1}\right), \tag{5.5}$$

$i_j=1,\ldots,p_j$, $j=1,\ldots,d-1$.

Since we have to multiply the factor by $\frac{\mathbb{L}^{-1}}{1-\mathbb{L}^{-1}}$ and $\mathbb{L}^{-1}=a_d$, then the product $(1-a_d^{n_1}a_{d+1})\frac{\mathbb{L}^{-1}\Phi_{\Delta_\gamma^t}(\mathbf{a})}{(1-\mathbb{L}^{-1})}$ equals

$$\frac{a_d\sum_{i_1=1}^{p_1}\cdots\sum_{i_{d-1}=1}^{p_{d-1}}a_d^{-(\alpha+\sum_{j=1}^{d-1}i_jb_j)}\prod_{j=1}^{d-1}a_j^{i_j}}{(1-a_d)\prod_{j=1}^{d-1}(1-a_j^{p_j}a_{d+1}^{\bar{b}_j})}=\frac{a_d^{1-\alpha}\prod_{j=1}^{d-1}\sum_{i_j=1}^{p_j}(a_ja_d^{-b_j})^{i_j}}{(1-a_d)\prod_{j=1}^{d-1}(1-a_j^{p_j}a_{d+1}^{\bar{b}_j})}=$$

$$=\frac{a_d^{1-\alpha}\prod_{j=1}^{d-1}\left(a_ja_d^{-b_j}(1-(a_ja_d^{-b_j})^{p_j})\right)}{(1-a_d)\prod_{j=1}^{d-1}\left((1-a_ja_d^{-b_j})(1-a_j^{p_j}a_{d+1}^{\bar{b}_j})\right)}=\frac{a_d^{1-\alpha}\prod_{j=1}^{d-1}u_j}{(1-a_d)\prod_{j=1}^{d-1}(1-u_j)}. \tag{5.6}$$

The term $(1-a_d^{n_1}a_{d+1})\frac{\mathbb{L}^{-1}\Phi_{\Delta_\gamma}(\mathbf{a})}{(1-\mathbb{L}^{-1})}$ is obtained from equation (5.6) when $\alpha=0$.

For the vertex τ_0, after the simplification, $y_d^{n_1}y_{d+1}=\mathbb{L}^{-(n_1+1)}T^n=1$, $y_jy_d^{-b_j}=u_j$ and $y_j^{p_j}y_{d+1}^{\bar{b}_1}=w_j$, for $j=1,\ldots,d-1$. The term $A^t(\mathbf{y})$ multiplied by $1-\mathbb{L}^{-(n_1+1)}T^n=(1-y_d^{n_1}y_{d+1})$ is the same as (5.3) with a_i replaced by y_i. Therefore the term we are interested in is

$$(5.7)\qquad -\frac{T^{m\alpha}y_d^{-\alpha}\prod_{j=1}^{d-1}u_j}{(1-y_d)\prod_{j=1}^{d-1}(1-u_j)} = -\frac{a_d^{-\alpha}\prod_{j=1}^{d-1}u_j}{(1-y_d)\prod_{j=1}^{d-1}(1-u_j)},$$

because $T^{-1}y_d = a_d$. Furthermore $A(\mathbf{y})(1-\mathbb{L}^{-(n_1+1)}T^n)$ is obtained from equation (5.7) when $\alpha = 0$. Thus

$$A_1^t = (5.7) - (5.4) - (5.6) = -\frac{a_d^{-\alpha}y_d\prod_{j=1}^{d-1}u_j}{(1-y_d)\prod_{j=1}^{d-1}(1-u_j)}.$$

And substituting $\alpha = 0$ in the last identity we get A_1 :

$$(5.8)\qquad A_1 = -\frac{y_d\prod_{j=1}^{d-1}u_j}{(1-y_d)\prod_{j=1}^{d-1}(1-u_j)}.$$

Finally we compute the contribution of the B part in the arc decomposition for both power series. Let π_γ be the map defined by $x_l = \bar{x}_l^{p_l}$ and $z = (z_1+\bar{\beta})\prod_{l=1}^{d}\bar{x}_l^{\bar{b}_l}$ where $\bar{\beta}^{n_1} = \beta$. We compute the contribution using the Newton map π_γ associated with the unique Newton component of h and the corresponding $k[t]$-morphisms associated with each $\mathbf{g} \in G_\gamma$. For a pair (h,ω) in equation (4.21) we show that the contribution is

$$(5.9)\qquad \begin{aligned} &Z_{DL}(\bar{h},\bar{\omega},T) + \sum_{\mathbf{g}\in G_\gamma,\mathbf{g}\neq\mathbf{g}^1} \mathbb{L}^{-(\nu_{\mathbf{g}}+A_{\mathbf{g}})} Z_{DL}^{\mathcal{W}}(\tilde{h}_{\mathbf{g}}^t,\tilde{\omega}_{\mathbf{g}},T) + \\ &+ \sum_{\mathbf{g}\in G_\gamma,\mathbf{g}\neq\mathbf{g}^1} \mathbb{L}^{-(\nu_{\mathbf{g}}+A_{\mathbf{g}})} \sum_{\emptyset\subsetneq J\subset\{1,\ldots,d-1\}} (\mathbb{L}-1)^{\#J} Z_{DL}^{\mathcal{W}}((\tilde{h}_{\mathbf{g}}^t)_{(\mathbf{y},0)},(\tilde{\omega}_{\mathbf{g}})_{(\mathbf{y},0)},T), \end{aligned}$$

where the pair $(\bar{h},\bar{\omega})$ is the pull-back by π_γ of (h,ω), that is

$$\bar{h} = \bar{x}_1^{\tilde{N}_1}\ldots\bar{x}_d^{\tilde{N}_d}(z_1^{m_r}+\ldots)u(\bar{\mathbf{x}},z_1), \quad u(\mathbf{0},0)\neq 0,$$

with $\tilde{N}_l = p_lN_l + \bar{b}_ln$, $l = 1,\ldots,d$, and $\bar{\omega} = \bar{x}_1^{\tilde{\nu}_1-1}\ldots\bar{x}_d^{\tilde{\nu}_d-1}d\bar{\mathbf{x}}\wedge dz_1$.

For $(\tilde{h}_{\mathbf{g}}^t,\tilde{\omega}_{\mathbf{g}})$ the form $\tilde{\omega}_{\mathbf{g}}$ is $\bar{\omega}$. Fix $\mathbf{g}\in G_\gamma\backslash\{\mathbf{g}^1\}$. The map $\pi_\gamma\circ\pi_{\mathbf{g}}$ is $x_l = t^{\mu_l^{\mathbf{g}}p_l}\bar{x}_l^{p_l}$ and $z = t^{c^{\mathbf{g}}}(z_1+\bar{\beta})\prod_{l=1}^{d}\bar{x}_l^{\bar{b}_l}$. Then $\tilde{h}_{\mathbf{g}}^t = h\circ\pi_\gamma\circ\pi_{\mathbf{g}}$, that is

$$\tilde{h}_{\mathbf{g}}^t := t^{nc^{\mathbf{g}}+\sum_{l=1}^{d-1}\mu_l^{\mathbf{g}}p_lN_l}\bar{x}_1^{\tilde{N}_1}\ldots\bar{x}_d^{\tilde{N}_d}(z_1^{m_r}+\ldots)u(\bar{\mathbf{x}},z_1), \quad u(\mathbf{0},0)\neq 0.$$

For h^t the maps associated with the corresponding $k[t]$-morphisms associated with each $\mathbf{g}\in G_\gamma^t$ appear. Looking at equation (4.20) we have $r = 1, u(r) = 1$ and

$s_1^{1,1} = 1$. This shows that the contribution is

$$\sum_{\mathbf{g}\in G_\gamma^t} \mathbb{L}^{-(\nu_\mathbf{g}+A_\mathbf{g})} \Bigg(Z_{DL}^{\mathcal{W}}(\tilde{h}_{\gamma,\mathbf{g}}^t, \tilde{\omega}_{\gamma,\mathbf{g}}, T) + \tag{5.10}$$
$$+ \sum_{\emptyset\subsetneq J\subset\{1,\dots,d-1\}} (\mathbb{L}-1)^{\#J}\, Z_{DL}^{\mathcal{W}}((\tilde{h}_{\gamma,\mathbf{g}}^t)_{(\mathbf{y},0)}, (\tilde{\omega}_{\gamma,\mathbf{g}})_{(\mathbf{y},0)}, T) \Bigg).$$

In this case, once $\mathbf{g}\in G_\gamma^t$ is fixed, the mapping considered $\pi_\gamma\circ\pi_\mathbf{g}$ is defined by $x_l = t^{\mu_l^\mathbf{g} p_l}\bar{x}_l^{p_l}$ and $z = t^{c^\mathbf{g}}(z_1+\bar{\beta})\prod_{l=1}^d \bar{x}_l^{\bar{b}_l}$. The pull-back $\tilde{h}_{\gamma,\mathbf{g}}^t$ is

$$t^{nc^\mathbf{g}+\sum_{l=1}^{d-1}\mu_l^\mathbf{g} p_l N_l}\bar{x}_1^{\tilde{N}_1}\dots\bar{x}_d^{\tilde{N}_d}(z_1^{m_r}+\dots)u(\bar{\mathbf{x}},z_1), \quad u(\mathbf{0},0)\neq 0,$$

and $\tilde{\omega}_{\gamma,\mathbf{g}} = \bar{\omega}$.

Computations for (5.10) and for all terms but the first one in (5.9) are similar. In both cases, if either $\mathbf{g}\in G_\gamma^t$ or $\mathbf{g}\in G_\gamma\setminus\{\mathbf{g}^1\}$ are fixed, then the local contribution from $Z_{DL}^{\mathcal{W}}(\tilde{h}_{\gamma,\mathbf{g}}^t, \tilde{\omega}_{\gamma,\mathbf{g}}, T)$ or from $Z_{DL}^{\mathcal{W}}(\tilde{h}_\mathbf{g}^t, \tilde{\omega}_\mathbf{g}, T)$ is given by the z_1-highest vertex $\tilde{\tau}$ which corresponds in both cases to the monomial

$$t^{\sum_{l=1}^{d-1}\mu_l^\mathbf{g} p_l N_l + nc^\mathbf{g}}\bar{x}_1^{\tilde{N}_1}\dots\bar{x}_d^{\tilde{N}_d} z_1^{m_r}.$$

We showed in Step 6 of the proof of theorems 4.3 and 4.9 that the factor coming from $\tilde{\tau}$ is $T^{\sum_{l=1}^{d-1}\mu_l^\mathbf{g} p_l N_l + nc^\mathbf{g}}\mathbb{L}^{-(d+1)}(\mathbb{L}-1)^{d+1}\Phi_{\Delta_{\tilde{\tau}}}(\mathbf{z})$, where $\mathbf{z}$ is equal to

$$\mathbf{z} := (\mathbb{L}^{-(\nu_1 p_1+\bar{b}_1)}T^{N_1p_1+n\bar{b}_1}, \dots, \mathbb{L}^{-(\nu_{d-1}p_{d-1}+\bar{b}_{d-1})}T^{N_{d-1}p_{d-1}+n\bar{b}_{d-1}}, \mathbb{L}^{-(n_1+1)}T^n, \mathbb{L}^{-1}T^m).$$

Since the corresponding cone $\Delta_{\tilde{\tau}}^c$ is the positive cone $\mathbb{P}^{d+1}$ (whose fundamental set $\tilde{G}$ has only one element $\tilde{G}=\{\mathbf{1}\}$), its contribution is

$$T^{\sum\limits_{l=1}^{d-1}\mu_l^\mathbf{g} p_l N_l + nc^\mathbf{g}}\mathbb{L}^{-(d+1)}(\mathbb{L}-1)^{d+1}S^1(\mathbf{z}),$$

where

$$S^1(\mathbf{x}) := \frac{\mathbf{x}^\mathbf{1}}{\prod\limits_{j=1}^{d+1}(1-x_j)} = \frac{x_1x_2\dots x_{d+1}}{\prod\limits_{j=1}^{d+1}(1-x_j)}. \tag{5.11}$$

Take any $\emptyset\subsetneq J\subsetneq\{1,\dots,d-1\}$. The number e of essential variables of $(\tilde{h}_{\gamma,\mathbf{g}}^t)_{(\mathbf{y},0)}$ and of $(\tilde{h}_\mathbf{g}^t)_{(\mathbf{y},0)}$ is $e = d+1-\#J$. By Corollary 5.1, $Z_{DL}^{\mathcal{W}}((\tilde{h}_{\gamma,\mathbf{g}}^t)_{(\mathbf{y},0)}, (\tilde{\omega}_{\gamma,\mathbf{g}})_{(\mathbf{y},0)}, T)$ is $\mathbb{L}^{-\#J}(\mathbb{L}-1)^e\mathbb{L}^{-e}T^{\sum_{l=1}^{d-1}\mu_l^\mathbf{g} p_l N_l + nc^\mathbf{g}}S_J^1(\mathbf{z})$ where $S_J^1(\mathbf{x}) := \frac{1}{(1-x_{d+1})}\prod_{j\notin J}\frac{x_j}{(1-x_j)}$. And the same identity is true for the zeta function $Z_{DL}^{\mathcal{W}}((\tilde{h}_\mathbf{g}^t)_{(\mathbf{y},0)}, (\tilde{\omega}_\mathbf{g})_{(\mathbf{y},0)}, T)$.

Thus we get all contributions we are interested in for (5.10) adding all previous results:

$$\mathbb{L}^{-(d+1)}(\mathbb{L}-1)^{d+1}\sum_{\mathbf{g}\in G_\gamma^t}\mathbb{L}^{-(\nu_\mathbf{g}+A_\mathbf{g})}T^{\sum_{l=1}^{d-1}\mu_l^\mathbf{g} p_l N_l + nc^\mathbf{g}}\left(S^1(\mathbf{z}) + \sum_{\emptyset\subsetneq J\subset\{1,\dots,d-1\}} S_J^1(\mathbf{z})\right).$$

The formula $S^1(\mathbf{x}) + \sum_{\emptyset \subsetneq J \subset \{1,\dots,d-1\}} S^1_J(\mathbf{x}) = S(\mathbf{x})$ is proved by induction where $S(\mathbf{x}) := \frac{x_d x_{d+1}}{\prod_{l=1}^{d+1}(1-x_l)}$. Thus we must simplify the contribution

$$B_1^t := (1 - \mathbb{L}^{-(n_1+1)}T^n)\left(\sum_{\mathbf{g}\in G_\gamma^t} \mathbb{L}^{-(\nu_{\mathbf{g}}+A_{\mathbf{g}})}T^{\sum_{l=1}^{d-1}\mu_l^{\mathbf{g}} p_l N_l + nc^{\mathbf{g}}} S(\mathbf{z})\right)$$

under our hypothesis $\mathbb{L}^{-(n_1+1)}T^n = 1$.

In the other case, equation (5.9), the contribution of $Z_{DL}(\bar{h}, \bar{\omega}, T)$ comes from the z_1-highest vertex $\tilde{\tau}$ which corresponds to the monomial $\bar{x}_1^{\tilde{N}_1} \dots \bar{x}_d^{\tilde{N}_d} z_1^{m_r}$. Thus it comes from $\mathbb{L}^{-(d+1)}(\mathbb{L}-1)^{d+1}\Phi_{\Delta_{\tilde{\tau}}}(\mathbf{z})$, where $(\mathbf{z})$ was defined before. In fact since the corresponding cone $\Delta^c_{\tilde{\tau}}$ is the positive cone $\mathbb{P}^{d+1}$ then the contribution is $\mathbb{L}^{-(d+1)}(\mathbb{L}-1)^{d+1}S^1(\mathbf{z})$ where $S^1(\mathbf{x})$ is given by equation (5.11). Thus all contributions we get when the Newton maps of (5.9) are done are

$$\mathbb{L}^{-(d+1)}(\mathbb{L}-1)^{d+1}\left(S^1(\mathbf{z}) + \sum_{\mathbf{g}\in G_\gamma, \mathbf{g}\neq \mathbf{g}^1} \mathbb{L}^{-(\nu_{\mathbf{g}}+A_{\mathbf{g}})}T^{\sum_{l=1}^{d-1}\mu_l^{\mathbf{g}} p_l N_l + nc^{\mathbf{g}}} S(\mathbf{z})\right).$$

We should simplify under our hypothesis $\mathbb{L}^{-(n_1+1)}T^n = 1$ the formula

$$B_1 := (1 - \mathbb{L}^{-(n_1+1)}T^n)\left(S^1(\mathbf{z}) + \sum_{\mathbf{g}\in G_\gamma, \mathbf{g}\neq \mathbf{g}^1} \mathbb{L}^{-(\nu_{\mathbf{g}}+A_{\mathbf{g}})}T^{\sum_{l=1}^{d-1}\mu_l^{\mathbf{g}} p_l N_l + nc^{\mathbf{g}}} S(\mathbf{z})\right).$$

Recall that the $(d+1)$-tuple $\mathbf{z}$ is $(w_1, \dots, w_{d-1}, 1, y_d)$.

Let us study the case B_1^t. For every $\mathbf{g}$ parametrized by $(i_1, \dots, i_{d-1})$ from equation (5.5) we have the following identity, which is only valid after the simplification below:

$$\nu_{\mathbf{g}} + A_{\mathbf{g}} = \frac{\alpha}{n_1} + \sum_{j=1}^{d-1}\frac{i_j}{p_j}(p_j\nu_j + \bar{b}_j) - \frac{\alpha + \sum_{j=1}^{d-1} i_j b_j}{n_1}(n_1+1).$$

It turns out that $\nu_{\mathbf{g}} + A_{\mathbf{g}} = -\alpha - \sum_{j=1}^{d-1} i_j(b_j - \nu_j)$. Since $c^{\mathbf{g}} = \frac{\alpha}{n_1} + \sum_{l=1}^{d}\mu_l\bar{b}_l$ then, under our simplification, $c^{\mathbf{g}} = 0$ and the exponent of T is $\sum_{j=1}^{d-1} i_j N_j$.

Therefore B_1^t equals:

$$\frac{\mathbb{L}^\alpha z_{d+1}\prod_{j=1}^{d-1}\sum_{k=1}^{p_j}(\mathbb{L}^{(b_j-\nu_j)}T^{N_j})^k}{(1-z_{d+1})\prod_{j=1}^{d-1}(1-z_j)} = \frac{\mathbb{L}^\alpha z_{d+1}\prod_{j=1}^{d-1}\left(u_j(1-u_j^{p_j})\right)}{(1-z_{d+1})\prod_{j=1}^{d-1}((1-z_j)(1-u_j))}.$$

Written as before we have

$$B_1^t = \frac{a_d^{-\alpha} y_d \prod_{j=1}^{d-1} u_j}{(1-y_d)\prod_{j=1}^{d-1}(1-u_j)}. \tag{5.12}$$

Thus

$$A_1^t + B_1^t = (5.7) - (5.4) - (5.6) + (5.12) = 0.$$

Let us study the case B_1. For every $\mathbf{g} \neq \mathbf{g}^1$ parametrized by $(i_1, \ldots, i_{d-1})$ from equation (5.5) with $i_j \in \{1, \ldots, p_j - 1\}$, we have $\nu_{\mathbf{g}} + A_{\mathbf{g}} = -\sum_{j=1}^{d-1} i_j(b_j - \nu_j)$. In the same way the exponent T is $\sum_{j=1}^{d-1} i_j N_j$.

The first term of B_1 is

$$(1 - z_d)S^1(\mathbf{z}) = \frac{y_d \prod_{j=1}^{d-1} w_j}{(1 - y_d)\prod_{j=1}^{d-1}(1 - w_j)} = \frac{y_d \prod_{j=1}^{d-1} u_j^{p_j}}{(1 - y_d)\prod_{j=1}^{d-1}(1 - u_j^{p_j})}. \tag{5.13}$$

The other terms come from

$$\frac{z_{d+1}}{(1 - z_{d+1})\prod_{j=1}^{d-1}(1 - z_j)} \prod_{j=1}^{d-1}\sum_{k=1}^{p_j - 1}\left(\mathbb{L}^{(b_j - \nu_j)}T^{N_j}\right)^k.$$

This is

$$\frac{y_d}{(1 - y_d)\prod_{j=1}^{d-1}(1 - u_j^{p_j})} \prod_{j=1}^{d-1}\sum_{k=1}^{p_j - 1}(u_j)^k = \frac{y_d}{(1 - y_d)\prod_{j=1}^{d-1}(1 - u_j^{p_j})} \prod_{j=1}^{d-1} u_j \left(\frac{1 - u_j^{p_j}}{1 - u} - u_j^{p_j - 1}\right). \tag{5.14}$$

Thus $A_1 + B_1 = (5.8) + (5.13) + (5.14) = 0$.

□

Let $h \in k[[\mathbf{x}]][z]$ be a quasi-ordinary power series in good coordinates with $h(\mathbf{x}, z) = x_1^{N_1} x_2^{N_2} \ldots x_d^{N_d} g(\mathbf{x}, z)$, where no x_i divides $g(\mathbf{x}, z)$ and $N_l \geq 0$ for any $l = 1, \ldots, d$. Write $h(\mathbf{x}, z) = x_1^{N_1} x_2^{N_2} \ldots x_d^{N_d} f(\mathbf{x}, z)u(\mathbf{x}, z)$, where $f(\mathbf{x}, z)$ is a quasi-ordinary z-polynomial of degree n in $k[[\mathbf{x}]][z]$. Let $\omega = \prod_{j=1}^{d} x_j^{\nu_j - 1} dx_1 \wedge \ldots \wedge dx_d \wedge dz$, $\nu_j \geq 1$ define a form such that (h, ω) satisfies the support condition (1.3). The set $CP(h, \omega)$ of *candidate poles* was defined in (3.28). Proposition 5.4 will allow us to consider a smaller set of candidate poles than the set $CP(h, \omega)$.

DEFINITION 5.5. A compact 1-dimensional face γ_q in $ND(h)$ will be called *special* in the i-th coordinate if the pair (h_{γ_q}, ω) satisfies the conditions of Proposition 5.4 in the i-th coordinate (instead of d-th coordinate). The face γ_q will be called *special* if it is special in one of the coordinates. If γ_q is special in the i-th and j-th, $(i \neq j)$, coordinates then the corresponding candidate poles coincide. Define

$$CP(h, \omega)_q := \left\{\left(\frac{M_l^q}{c_l^q}, \nu_l p_l^q + \bar{b}_l^q\right)\right\}_{l=1,\ldots,\widehat{i},\ldots,d} \quad \text{if } \gamma_q \text{ is special,}$$

where we assume that γ_q is special in the i-th coordinate and $\widehat{i}$ means that we omit i. In particular, if γ_q is special in more than one variable then the corresponding pair (N, ν) appears in $CP(h, \omega)_q$. Otherwise, define

$$CP(h,\omega)_q := \left\{ \left(\frac{M_l^q}{c_l^q}, \nu_l p_l^q + \bar{b}_l^q \right) \right\}_{l=1}^{d} \quad \text{if } \gamma_q \text{ is not special.}$$

Let us define by induction the sets

$$\widetilde{CP}(h,\omega) := \bigcup_{q=1}^{r} CP(h,\omega)_q \cup \bigcup \widetilde{CP}(\bar{h}_{q,j}, \bar{\omega}_{q,j})$$

where the union is over all pull-back $(\bar{h}_{q,j}, \bar{\omega}_{q,j})$ under the Newton maps associated with Newton components f_j^q of f. The set of *strong candidate poles* of a pair (h, ω) is defined by

$$SCP(h,\omega) := \{(N_i, \nu_i)\}_{i=1}^{d} \cup \widetilde{CP}(h,\omega).$$

After Proposition 5.4 and Theorem 5.3 the following result is clear.

PROPOSITION 5.6. *If $Ns + \nu$ is the equation of a pole of $Z_{top,0}(h, \omega, s)$ then $(N, \nu) \in SCP(h, \omega)$.*

In fact Proposition 5.4 and equations (4.20) and (4.21) show that the same result is true for the local Denef-Loeser motivic zeta function.

PROPOSITION 5.7.

$$Z_{DL}(h,\omega,T) \in \mathbb{Z}[\mathbb{L}, \mathbb{L}^{-1}, (1-\mathbb{L}^{-\nu}T^N)^{-1}][T]_{(N,\nu)\in SCP(h,\omega)}.$$

REMARK 5.8. Let $h = \prod_{l=1} x_l^{N_l} f(\mathbf{x}, z) u(\mathbf{x}, z)$, with $u(\mathbf{0}, 0) \neq 0$. Take any compact 1-dimensional face γ of $\Gamma(h)$. For each $j \in \{1, \ldots, d\}$ such that γ is not contained in the (x_j, z)-plane, consider the j-transversal section h_j^0 with root $\alpha = 0$. Let $\tilde{\gamma}$ be the compact face of $\Gamma(h_j^0)$ on which γ is projected. The face $\tilde{\gamma}$ can be the projection of several distinct compact faces of $\Gamma(h)$. Thus if γ is not special then its projection $\tilde{\gamma}$ cannot be special in any coordinate.

STRONG CANDIDATE POLES FOR $\mathbf{ev}(h) = 1$. Assume $h = x_1^{N_1} f(x_1, z) u(x_1, z) \in k[[x_1]][z]$ is a quasi-ordinary power series in good coordinates with $\mathbf{ev}(h) = 1$. Let γ be a compact 1-dim face of $\Gamma(h)$. Then γ is special if and only if γ is the highest (with respect to z) compact face of $\Gamma(h)$ and h_γ is of type $z^{n-n_1 m}(z^{n_1} - \beta x_1)^{n-n_1}$, $\beta \in \mathbf{G}_{m,k}$. In particular $N_1 = 0$. We observe that γ is special if and only we can apply the inversion formula to h permuting coordinates. If γ is special then $CP(h, \omega)_\gamma = \emptyset$, for each pair (h, ω) satisfying the support condition (1.3). After a Newton map associated with a Newton component of f, the pull-back of h is of type $y^a g(y, z_1)$ with $a > 0$. In particular Proposition 5.4 cannot be applied anymore.

If $\mathbf{ev}(h) > 1$, more general facts occur. Take for instance any of the following examples: $f(x, y, z) = (z^2 - x^3)^2 + x^{11} y$ and $g(x, y, u, z) = ((z^2 - x^3 y)^2 + x^7 y^2)((z^2 - x^3 y u)^2 + x^7 y^2 u^3)$. For f, after the unique possible Newton map π, a special face for $f \circ \pi$ appears. In the other case, $\Gamma(g)$ has two compact 1-dimensional faces and both are special, in different coordinates.

STRONG CANDIDATE POLES FOR $\mathbf{ev}(h) = 2$. Let $h \in k[[x_1, x_2]][z]$ be a quasi-ordinary power series in good coordinates with $\mathbf{ev}(h) = 2$. Assume that we can decompose $h = x_1^{N_1} x_2^{N_2} f(x_1, x_2, z) u(x_1, x_2, z)$, with $u(\mathbf{0}) \neq 0$ and f is a Weierstrass polynomial of degree n. Let γ denote the highest (with respect to z) compact face of $\Gamma(h)$ and let $h_\gamma = x_1^{N_1}x_2^{N_2}z^{n-n_1m}\prod_{j=1}^{u}(z^{n_1}-\beta_j x_1^{b_1}x_2^{b_2})^{m_j}$ with $\beta_j \in \mathbf{G}_{m,k}$ and $m=\sum m_j$. Let $\omega = x_1^{\nu_1-1}x_2^{\nu_2-1}dx_1\wedge dx_2\wedge dz$ be a differential form such that (h,ω) satisfies (1.3).

LEMMA 5.9. *$(n, n_1+1) \in CP(h,w)_\gamma$ if and only if either*

(1) *there exists $i\in\{1,2\}$ such that $b_i=1$, $N_i=0$ (this implies $\nu_i=1$) and $u>1$, or*

(2) *$u=1$ and $b_1=b_2=1$, $N_1=N_2=0$ (this implies $\nu_1=\nu_2=1$) and therefore γ is special in both coordinates.*

PROOF. By definitions and Lemma 3.15, γ is defined by $n_1x_1+b_1z = n_1N_1+b_1n$ and $n_1x_2+b_2z=n_2N_2+b_2n$. The candidate poles from γ are $\left(\frac{n_1N_i+b_in}{\gcd(n_1,b_i)}, \frac{n_1\nu_i+b_i}{\gcd(n_1,b_i)}\right)$ with $i=1,2$. In particular $\left(\frac{n_1N_i+b_in}{\gcd(n_1,b_i)}, \frac{n_1\nu_i+b_i}{\gcd(n_1,b_i)}\right) = (n,n_1+1)$ if and only if either $b_i=0$ and then $\gcd(n_1,b_i)=n_1$, which implies $(N_i,\nu_i)=(n,n_1+1)\in CP(h,w)_\gamma$ which is absurd. Or one of the two conditions of the lemma is satisfied. □

In case (1), since $u>1$, (n,n_1+1) is a strong candidate pole for h_j^α, $j\neq i$). In case (2) the i-transversal section h_i^α of h has only $\alpha=0$ as a root because $(\frac{1}{n},\frac{1}{n})$ is the smallest characteristic exponent of f, it is the z-highest face of $NP(h)$. In particular h_i^0 has at least one Newton component associated with the projection $\tilde{\gamma}$ of γ on the plane x_iz. Thus (n,n_1+1) is a candidate pole for (h_i^0,ω_i). In fact, $\tilde{\gamma}$ is the highest face with respect to z. Applying the case $\mathbf{ev}(h)=1$ then (n,n_1+1) is not a strong candidate pole for both h_i^0.

PROPOSITION 5.10. *Under the above conditions one has*

$$SCP(h,\omega)\subset \bigcup_{i=1}^{2}\bigcup_{m=1}^{v_i} SCP(h_i^{\alpha_m},\omega_i)\cup\{(n,n_1+1)\},$$

where $h_i^{\alpha_m}$ are the corresponding i-transversal section at the root α_m, see inclusion (3.19) *in Proposition 3.40.*

PROOF. Let us prove by induction on depth(h), that if we are in good coordinates, with the above definitions of n, n_1 and γ then

$$SCP(h,\omega)\subset \bigcup_{i=1,2}\bigcup_{m=1}^{v_i} SCP(h_i^{\alpha_m},\omega_i)\cup\{(n(h),n_1(h)+1)\}.$$

We write $(n(h),n_1(h)+1)$ to emphasize that at each step of the induction they depend on h.

If depth$(h)=0$ then $h=x_1^{N_1}x_2^{N_2}z\,u(\mathbf{x},z)$, with $u(\mathbf{0})\neq 0$ The result follows easily. Assume that we have proved the result for depth$(h)<m$. Let us decompose $h=x_1^{N_1}x_2^{N_2}f(x_1,x_2,z)u(x_1,x_2,z)$, with $u(\mathbf{0})\neq 0$ and depth$(h)=m$. Let $(N,\nu)\in SCP(h,\omega)\subset CP(h,\omega)$. If $(N,\nu)=(n,n_1+1)$ we are done. If $(N,\nu)=(N_i,\nu_i)\neq(0,1)$, $i=1,2$ then any of the factors h_j^α of the j-transversal section with $i\neq j$ has $x_i^{N_i}$ as a factor which implies $(N,\nu)=(N_i,\nu_i)\in SCP(h_j^\alpha,\omega_j)$.

If $(N,\nu)\in\bigcup_{q=1}^{r}CP(h,\omega)_q$ where $\Gamma(h)$ has r compact 1-dim faces. Assume (N,ν) appears in the compact face γ_q, say in the i-th coordinate. Take the transversal section h_j^0, $j\neq i$. Let $\tilde{\gamma}$ be the compact face of $\Gamma(h_j^0)$ on which γ_q is projected.

If γ_q is contained in the (x_i, z)-plane, then $\tilde{\gamma}$ contains γ_q. If we can apply proposition 5.4 to h_j^0 and $\tilde{\gamma}$ then $(h_j^0)|_{\tilde{\gamma}} = z^{n-mn_q}(z^{n_q} - \beta x_i)^m, \beta \in \mathbf{G}_{m,k_j}$. Then γ_q must corresponds to the smallest characteristic exponent of h. Thus γ_q is special in this coordinate. This contradicts $(N, \nu) \in SCP(h, \omega)$.

Otherwise, Remark 5.8 contradicts $(N, \nu) \in CP(h, \omega)_q$ if γ_q is not special in the i-th coordinate. If γ_q is special in the i-th coordinate but not in the j-coordinate then $(N, \nu) \notin CP(h, \omega)_q$ which is absurd. Therefore if γ_q is special in both coordinates then γ_q is the highest compact face and by Lemma 5.9(2), $(N, \nu) = (n, n_1 + 1)$. Finally we apply induction.

To finish the proof of the proposition is enough to remark that the conditions in Lemma 5.9(2) cannot be obtained after any Newton map. The reason for that is that after any Newton map the pull-back of h has at least a factor of type $y_i^{N_i}$ with $N_i > 0$. Therefore the proof is finished. □

STRONG CANDIDATE POLES FOR $\mathbf{ev}(h) = d > 2$. Let $h \in k[[x_1, \ldots, x_d]][z]$ be a quasi-ordinary power series in good coordinates with $\mathbf{ev}(h) = d > 2$. Assume that $h = \prod_{l=1}^{d} x_l^{N_l} f(\mathbf{x}, z) u(\mathbf{x}, z)$, with $u(\mathbf{0}) \neq 0$ and f is a Weierstrass polynomial of degree n. Let ω be a differential form such that (h, ω) satisfies the support condition 1.3.

PROPOSITION 5.11. *Under the above conditions one has*

$$SCP(h, \omega) \subset \bigcup_{i=1,\ldots,d,} \bigcup_{m=1}^{v_i} SCP(h_i^{\alpha_m}, \omega_i),$$

where $h_i^{\alpha_m}$ are the corresponding i-transversal section at the root α_m, see lemma 3.19.

PROOF. The proof is by induction on the depth(h) and follows the same ideas as proof of proposition 5.10 but in this case we do not need to take care of the special compact 1-dimensional faces. □

COMPLEX ANALYTIC SET UP. If $k = \mathbb{C}$ and we work with convergent complex quasi-ordinary power series $h \in \mathbb{C}\{\mathbf{x}\}[z]$ all results presented in the last three sections are valid too. We leave the details to the reader. The pull-back under the Newton maps of convergent quasi-ordinary power series are again convergent. The transversal sections can be seen now as follows. We write $h(\mathbf{x}, z) = \prod_{l=1}^{d} x_l^{N_l} f(\mathbf{x}, z) u(\mathbf{x}, z)$, where no x_i divides $f(\mathbf{x}, z)$, $u(0, 0) \neq 0$, $N_l \geq 0$ for any $l = 1, \ldots, d$, and f is a Weierstrass polynomial in z. After Lemma 5.1 we may assume that $\mathbf{ev}(h) = d$. We may assume f is convergent in a polydisk $\Delta_\varepsilon^d \times \Delta_\delta^1$, where $0 \ll \delta \ll \varepsilon \ll 1$ and if $\mathbf{x}^0 \in \Delta_\varepsilon^d$ all roots of $f(\mathbf{x}^0, z)$ lie in Δ_δ^1.

Let us consider the germ $(\text{Sing}(V), 0)$ of singular points of $V = h^{-1}(0)$ in a neighborhood of the origin. The condition of the discriminant implies that $(\text{Sing}(V), 0)$ is contained in the intersection of $h^{-1}(0)$ with the hyperplanes $x_l = 0$, $1 \leq l \leq d$. Fix $i \in \{1, \ldots, d\}$. Consider the polynomial, over $\mathbb{C}\{x_i\}$, of degree n

$$f(\mathbf{0}_i(x_i), z) = z^n + \text{ lower degree terms.}$$

where the d-tuple $\mathbf{0}_i$ has all coordinates 0 but the i-th coordinate which is x_i. We have two possibilities:

(1) If $f(\mathbf{0}_i(x_i), z) = z^n$ then the points $(\mathbf{0}_i(x^0), 0)$ belong to $h^{-1}(0)$, for all $x^0 \in \Delta^1_\varepsilon$. The power series $h_i^{0,x^0}(\mathbf{x}, z) := h(\mathbf{x} + \mathbf{0}_i(x^0), z)$ are quasi-ordinaries. They are the analytic equivalent to the formal i-transversal section at the root $\alpha = 0$.

(2) Otherwise, i.e. if $f(\mathbf{0}_i(x_i), z) \neq z^n$, given $x^0 \in \Delta^1_\varepsilon$ there exist $z^0_1, \ldots, z^0_l \in \Delta^1_\delta$, pairwise distinct, and $r_1, \ldots, r_l \in \mathbb{P}$, $\sum r_j = n$, such that

$$f(\mathbf{0}_i(x^0), z) = \prod_{j=1}^{l} (z - z^0_j)^{r_j}.$$

We can suppose that ε is small enough in order to have that $r_1, \ldots, r_l$ independent of x^0. Again $h_i^{\alpha,x^0}(\mathbf{x}, z) := h(\mathbf{x} + \mathbf{0}_i(x^0), z + z^0_j)$ are quasi-ordinary power series. In this case they are the equivalent in the analytic case to the i-transversal at some root $\alpha \neq 0$.

In particular we have the analytic analogue of proposition 3.40.

CHAPTER 6

Monodromy conjecture for quasi-ordinary power series

Let $h : (\mathbb{C}^{d+1}, 0) \to (\mathbb{C}, 0)$ be a germ of complex analytic function such that $h(0) = 0$. Fix U a sufficiently small neighborhood of 0 where h is defined. Let F be the Milnor fiber of the Milnor fibration at the origin associated with h. Let $m_F : F \to F$ be the monodromy transformation. The zeta-function of the monodromy of h is

$$\zeta(h)(t) := \prod_{q \geq 0} \det(I - t \quad m_{F\,q})^{(-1)^q}$$

where $m_{F\,q} : H_q(F, \mathbb{C}) \to H_q(F, \mathbb{C})$, $q \geq 0$, are the homological monodromy transformations.

In this section we solve in the quasi-ordinary case the monodromy conjectures stated in the Introduction. We will focus firstly on the the topological monodromy conjecture of [**14**] and on the *motivic monodromy conjecture* of [**15**, section 2.4] as stated in the introduction but for convergent quasi-ordinary series.

The monodromy conjectures deal with eigenvalues of the complex algebraic monodromy at some points of the zero locus of h. First we will prove the monodromy conjecture for $\mathbf{ev}(h) = 1$. As we mention in the introduction the result is known (several proofs by Loeser, Veys) but we present here an independent proof following our ideas which by the way will be useful for the proof in the general case. In the general case, i.e $\mathbf{ev}(h) > 1$, we need to consider the monodromy at some different points of the singular locus of h. These points will be at some transversal sections of h. Thus we proceed by induction on $\mathbf{ev}(h)$.

1. Monodromy conjecture for curves

Recall that any germ $h \in \mathbb{C}\{x, z\}$ of a curve is quasi-ordinary. We will prove that each Newton process produces eigenvalues of the zeta function of the monodromy of the curve defined by h. This is true in all cases except for the *special candidate pole* but the special pole candidate (if it appears) is not a pole for the motivic zeta function (hence for the topological zeta function) because of Proposition 5.4.

THEOREM 6.1. *Let $h(x, z) \in \mathbb{C}\{x\}[z]$ be a germ of a curve and ω a regular differential form such that (h, ω) satisfies the support condition (1.3). If $(N, \nu) \in SCP(h, \omega)$ let $q := -\frac{\nu}{N}$, then $\exp(2i\pi q)$ is an eigenvalue of the zeta function of the monodromy at some point of $h^{-1}(0)$.*

PROOF. The case $\mathrm{depth}(h) = 0$ is trivial, since $h = x^N z$. In this case $SCP(h) = \{(N, \nu), (1, 1)\}$, where ν comes from the differential form; at some points of $h^{-1}(0)$, the zeta function of the monodromy is either $1 - t^N$ or $1 - t$ and the result follows.

We prove by induction on the depth ≥ 1 that if $(N,\nu) \in SCP(h,\omega)$ and $q := -\frac{\nu}{N}$, then $\exp(2i\pi q)$ is an eigenvalue of the zeta function of the monodromy. We will not consider the case $q \in \mathbb{Z}$ since this case always gives the eigenvalue 1 of the monodromy.

Let h be a curve singularity with depth $m+1$ in good coordinates, $m \geq 0$. Assume $h = x^N z^\varepsilon g(x,z)$, $\varepsilon = 0,1$ and $\omega = x^{\nu-1}dx \wedge dz$, where neither x nor z divide g and (h,ω) satisfies the support condition 1.3. Let $\gamma_1,\ldots,\gamma_r$ be the compact edges of the Newton diagram of h; for each $q = 1,\ldots,r$, we have denoted by $v(q)$ the number of non-zero distinct roots of h_{γ_q}. Each one of these roots defines a Newton map and let $h_{q,j}$ $j = 1,\ldots,v(q)$, be the pull-backs of h by these Newtons maps. Recall that the vertex τ_r of γ_r is the z-highest vertex. We will define also α,β such that $\alpha = 0$ (resp. $\beta = 0$) if the slope of γ_r (resp. the inverse of the slope of γ_1) is an integer and x (resp. z) does not divide h. Otherwise we set $\alpha = 1$ (resp. $\beta = 1$). Besides the candidate poles (N,ν) and $(1,1)$, denote by (N_q,ν_q), $q = 1,\ldots,r$, the candidate poles corresponding to the edges. It is clear that with the poles (N,ν) and $(1,1)$ we can complete our argument as above. Note that:

- $\operatorname{depth}(h) = 1$ if and only if $\operatorname{depth}(h_{q,j}) = 0$; the corresponding monodromy zeta functions for these pull-backs are equal to 1.
- h is in good coordinates if and only if $v(1) + \beta > 1$.
- (N_r,ν_r) is a strong candidate pole if and only if $v(r) + \alpha > 1$; recall that (N_q,ν_q), $q = 1,\ldots,r-1$, are strong candidate poles.

One can prove, see e.g. [**7**], that at the corresponding points of the strict transform of h after Newton mappings and after the partial resolution induced by the Newton polygon of h the total transforms of h are isomorphic curves. Using A'Campo's formula we deduce that the monodromy zeta function of h at the origin is

$$\zeta(h)(t) = \frac{(1-t^{N'_r})^a}{(1-t^{N_r})^{\alpha-1}}\frac{(1-t^{N''_1})^b}{(1-t^{N_1})^{\beta-1}}\prod_{q=1}^{r}\left((1-t^{N_q})^{-v(q)}\prod_{j=1}^{v(q)}\zeta(h_{q,j})(t)\right), \tag{6.1}$$

where

- N''_1 divides N_1, N'_r divides N_r.
- If x (resp. z) divides h then $a = 0$ (resp. $b = 0$) and the corresponding numerator does not appear, otherwise $a = 1$ (resp. $b = 1$).
- If x does not divide h, this implies $(N,\nu) = (0,1)$ and we can express $h_{\gamma_r} = z^{u_r}\tilde{h}_r$ where $\tilde{h}_r$ is a product of $m_r = \sum_{j=1}^{v(r)} m_{r,j}$ factors of the form $(z^{n_1^r} - vx^{b^r})$, counting multiplicities. Recall that in the case of curves we always have $\gcd(n_1^r, b^r) = 1$. Thus

$$N_r = (u_r + m_r n_1^r)b^r, \quad \nu_r = n_1^r + b^r, \quad N'_r = u_r + m_r n_1^r. \tag{6.2}$$

- If z does not divide h, that is $\varepsilon = 0$, we can express $h_{\gamma_1} = x^{u_1}\tilde{h}_1$ where $\tilde{h}_1$ is a product of m_1 factors of the form $(z^{n_1^1} - \tilde{v}x^{b^1})$, again $\gcd(n_1^1, b^1) = 1$. In this case

$$N_1 = (u_1 + m_1 b^1)n_1^1, \quad \nu_1 = \nu n_1^1 + b^1, \quad N''_1 = u_1 + m_1 b^1. \tag{6.3}$$

Recall that the inverse of the zeta function of the monodromy is, up to a factor of $t(1-t)$, a polynomial. In order to prove the theorem we must take special care of numerators in the formula (6.1). They can arise in several situations:

(1) $r > 1$, x does not divide h, $v(r) = 1$. The equations of (6.2) imply that $\exp(-2i\pi\frac{\nu_r}{N_r})$ is a root of the polynomial which is the inverse of the quotient $\frac{1-t^{N'_r}}{1-t^{N_r}}$.
(2) $r > 1$, z does not divide h, $v(1) = 1$. We proceed using equations (6.3).
(3) $r = 1$, either x or z divide h, $v(1) = 1$. We proceed using equations (6.2) or (6.3).
(4) $r = 1$, neither x nor z divide h, $v(1) = 1$. Using equations (6.2) it is easily seen that (N_1, ν_1) provides a root of the inverse of the monodromy zeta function. In such a case the zeta function is written as

$$\frac{(1-t^{N'_1})(1-t^{N''_1})}{(1-t^{N_1})}\zeta(h_{1,1})(t). \tag{6.4}$$

With these arguments, if $\operatorname{depth}(h) = 1$, the proof of the result is finished. Let us assume now that $\operatorname{depth}(h) > 1$. From the above computations we must prove that $SCP(h_{q,j})$ produce also roots of $(\zeta(h)(t))^{-1}$. By induction hypothesis this is true whenever $(\zeta(h_{q,j})(t))^{-1}$ divides $(\zeta(h)(t))^{-1}$. The unique case where this fact does not happen is in (6.4), since the first factor is not a polynomial. Let us rewrite the above formula using (6.2), (6.3) and $u_1 = u_r = 0$:

$$\frac{(1-t^{m_1 n_1^1})(1-t^{m_1 b^1})}{(1-t^{m_1 n_1^1 b^1})}\zeta(h_{1,1})(t). \tag{6.5}$$

Since the inverse of the zeta function of the monodromy is, up to a factor of $t(1-t)$, a polynomial, there must exist a factor $(1-t^{m_1})$ which divides $(\zeta(h_{1,1})(t))^{-1}$. Let us note that $h_{1,1} = x^{m_1 n_1^1 b^1}(z^{m_1}+\dots)$ and the new differential form is $x^{n_1^1+b^1-1}dx\wedge dz$. The highest vertex of the new Newton polygon is $(m_1 n_1^1 b^1, m_1)$; let us assume that we have s edges and denote $(\tilde{N}, \tilde{\nu})$ the strong candidate pole associated to highest edge. Let us denote $(\tilde{n}, \tilde{b})$ the coprime integers which provide the slope of the edge.

We have $\tilde{N} = m_1(\tilde{n}n_1^1 b^1 + \tilde{b})$, $\tilde{\nu} = \nu_1\tilde{n} + \tilde{b} = (n_1^1 + b^1)\tilde{n} + \tilde{b}$. The function $\zeta(h_{1,1})(t)$ is

$$\zeta(h_{1,1})(t) = (1-t^{\tilde{N}})^{-v(s)}\zeta(\bar{h}_{s,j})(t)\prod_{q=1}^{s-1}\Delta_q, \tag{6.6}$$

where Δ_q is the product of the factors associated to the $s-1$ first edges. The factor $(1-t^{\tilde{N}})$ is a multiple of $(1-t^{m_1})$. This means that $(\tilde{N}, \tilde{\nu})$ is the only candidate pole which may not give an eigenvalue of the monodromy of h. Since

$$(\tilde{N}, \tilde{\nu}) = (m_1(\tilde{n}n_1^1 b^1 + \tilde{b}), (n_1^1 + b^1)\tilde{n} + \tilde{b}),$$

it follows that $\exp(-2i\pi\frac{\tilde{\nu}}{\tilde{N}})^{m_1} = \exp(-2i\pi\frac{(n_1^1+b^1)\tilde{n}+\tilde{b}}{\tilde{n}n_1^1 b^1+\tilde{b}})$. It is easy to check that the rational number $\frac{(n_1^1+b^1)\tilde{n}+\tilde{b}}{\tilde{n}n_1^1 b^1+\tilde{b}}$ is not a positive integer. In particular $\exp(-2i\pi\frac{\tilde{\nu}}{\tilde{N}})^{m_1} \neq 1$ and it is an eigenvalue of the monodromy of h.. □

2. Monodromy conjecture: general case

We will need some facts about the zeta-function of the monodromy of quasi-ordinary analytic power series. A formula to compute the zeta function of the monodromy has been obtained by L.J. McEwan, A Némethi, and P.D. González Pérez, see [**38, 23**].

Assume h is a quasi-ordinary power series defined by $f(\mathbf{x}, z)u(\mathbf{x}, z)$ with $u(\mathbf{0}, 0) \neq 0$ and $f = z^n + a_1(\mathbf{x})z^{n-1} + \ldots$ is in good coordinates. If $\Lambda_{CE} = \emptyset$ then f is irreducible and smooth. Otherwise one reorders the variables $\mathbf{x}$ in such a way that the first entry of $\min \Lambda_{CE} \in \mathbb{Q}^d$ is non-zero. Since the set $\Lambda_{ND}(f) \subset \Lambda$ is totally ordered, $\min \Lambda_{CE}$ exists.

THEOREM 6.2. [**38, 23**] *With the above hypothesis, after the reordering of the coordinates described before one has* $\zeta(h)(t) = \zeta(f)(t) = \zeta\left(f|_{\{x_2=\ldots=x_d=0\}}\right)(t)$.

In fact, if $\min \Lambda_{CE}$ *has at least two non-zero entries, then* $\zeta(f)(t) = (1 - t^n)$.

COROLLARY 6.3. *If* $h \in \mathbb{C}\{\mathbf{x}\}[z]$ *is quasi-ordinary satisfying* $\mathbf{ev}(h) = 1$ *and* ω *a regular differential form such that* (h, ω) *satisfies the support condition (1.3), then the monodromy conjecture holds for* $Z_{DL}(h, \omega, T)$.

We can apply Theorem 6.2 and Corollary 5.1 to conclude the proof of Corollary 6.3 from Theorem 6.1.

EXAMPLE 6.4. If $f = z^n + x_1 \cdots x_r$. Then f has a non-degenerate Newton polyhedron. One can compute the topological zeta function (using Denef and Loeser formula or our algorithm) and the monodromy zeta function $\zeta_f(t)$ (using the above result or Varchenko formula, [**43**]). In fact $\zeta(f) = (1 - t^n)$ and $Z_{top,0}(f, s)$ has only two poles $s = -1$ and $s = -\frac{n+1}{n}$. In particular the monodromy conjecture holds.

THEOREM 6.5. *Let* $h(x, z) \in \mathbb{C}\{\mathbf{x}\}[z]$ *be a quasi-ordinary power series and* ω *a regular differential form such that* (h, ω) *satisfies the support condition 1.3. Then the monodromy conjecture holds for* $Z_{DL}(h, \omega, T)$ *and* $Z_{\text{top},0}(h, \omega, s)$.

PROOF. Let $h \in \mathbb{C}\{x_1, \ldots, x_d\}[z]$ be a quasi-ordinary analytic power series in good coordinates. We proceed by induction on $\mathbf{ev}(h) = d > 1$.

Assume $h \in \mathbb{C}\{x_1, x_2\}[z]$ has $\mathbf{ev}(h) = 2$. By Proposition 5.10, every strong candidate pole is either $(n, n_1 + 1)$ or it is a strong candidate pole of one of the transversal sections. In the former case, the highest compact 1-dimensional face of $\Gamma(h)$ is special in both coordinates. Thus $\min \Lambda_{CE}(h)$ has at least two non-zero entries, which implies $\zeta(h)(t) = (1 - t^n)$, after Theorem 6.2. In particular $\exp(-2i\pi(n_1 + 1)/n)$ is an eigenvalue of the monodromy of h at the origin.

Otherwise, since we have proved the result for $\mathbf{ev}(h) = 1$, $\exp(-2i\pi(n_1 + 1)/n)$ is an eigenvalue of the monodromy of one of the transversal sections at some point which is, in fact, contained in $h^{-1}(0)$.

Moreover for any quasi-ordinary analytic power series h such that $\mathbf{ev}(h) = 2$ the monodromy conjectures for motivic and topological zeta functions hold, after Corollary 5.1.

By Proposition 5.11 and the induction we can conclude that the monodromy conjecture for motivic and topological zeta functions hold for quasi-ordinary power series h such that $\mathbf{ev}(h) = d$. □

3. Monodromy conjecture for the Igusa zeta-function

Let p be a prime number and let K be a p-adic field, i.e $[K : \mathbb{Q}_p] < \infty$. Let R be the valuation ring of K, P the maximal ideal of R, and $\bar{K} = R/P$ the residue field of K. Let q denote the cardinality of $\bar{K}$, so $\bar{K} \simeq \mathbb{F}_q$. For z in K, let $\operatorname{ord} z$ denote the valuation of z, and set $|z| = q^{-\operatorname{ord} z}$. Let h be a non constant element in $K[x_1, \ldots, x_d]$. The p-adic *Igusa local zeta function* $I_0(h, K, s)$ associated with h (relative to the trivial multiplicative character) is defined as the p-adic integral

$$I_0(h, K, s) := \int_{PR^d} |h(x)|^s |dx|, \tag{6.7}$$

for $s \in \mathbb{C}$, $\Re(s) > 0$, where $|dx|$ denotes the Haar measure on K^d normalized in such of way that R^d is of volume 1. Igusa proved that $I_0(h, K, s)$ is a rational function of q^{-s}, see [**10**].

REMARK 6.6. The method described here can be used to compute $I_0(h, K, s)$ in the case of that $h \in K[x_1, \ldots, x_d, z]$ is a quasi-ordinary polynomial.

Assume now that h is a non-constant polynomial in $F[x_1, \ldots, x_d]$, for some number field $F \subset \mathbb{C}$. Igusa's monodromy conjecture states that for almost all p-adic completion K of F, if s_0 is a pole of $I_0(h, K, s)$, then $\exp(2i\pi\Re(s_0))$ is an eigenvalue of the local monodromy of h at some point of $h^{-1}(0)$.

It is known and deduced from [**15**], see also [**17**], that for almost all finite places of the number field F, the real parts $Ns+\nu$ of poles of $I_0(h, K, s)$ come from factors $(1-\mathbb{L}^{-\nu}T^N)$ in the denominator of $Z_{\text{naive}}(h, T)$ (which is essentially $Z_{\text{DL}}(h, T)$). In particular our proof can be applied to polynomials $h \in F[\mathbf{x}, z]$ with coefficients in a number field F which are quasi-ordinary polynomials. Then the Igusa monodromy conjecture is also true in this case.

Bibliography

[1] S.S. Abhyankar, *On the ramification of algebraic functions*, Amer. J. Math. **77** (1955), 575–592.

[2] E. Artal Bartolo, Pi. Cassou-Noguès, I. Luengo, A. Melle Hernández, *Monodromy conjecture for some surface singularities*, Ann. Scient. Éc. Norm. Sup. **35** (2002), 605–640.

[3] ———, *The Denef-Loeser zeta function is not a topological invariant*, J. London Math. Soc. (2) **65** (2002), no. 1, 45–54.

[4] ———, *Quasi-ordinary singularities from the Newton polyhedron point of view*, preprint 2004.

[5] A. Barvinok, J.E Pommersheim, *An algorithmic theory of lattice points in polyhedra*, New perspectives in algebraic combinatorics (Berkeley, CA, 1996–97), 91–147, Math. Sci. Res. Inst. Publ., **38**, Cambridge Univ. Press, Cambridge (1999), 91–147.

[6] D. Bernstein, A.G. Khovanskii and A.G. Kushnirenko, *Newton polyhedra*, Usp. Met. Nauk. **31** (1976), 201–202.

[7] Pi. Cassou-Noguès, A. Płoski, *Algebraic Curves*, Book in preparation.

[8] J. Cheah, *On the cohomology of Hilbert schemes of points*, J. Algebraic Geometry **5** (1996), 479–511.

[9] J. Denef, *The rationality of the Poincaré series associated to the p-adic points on a variety*, Invent. Math. **77** (1984), 1–23.

[10] ———, *Report on Igusa's local zeta function*, Séminaire Bourbaki, Astérisque **201-202-203** (1991), 358–386.

[11] ———, *Poles of p-adic complex powers and Newton polyhedra*, Nieuw Archief voor Wiskunde **13** (1995), 289–295.

[12] J. Denef, K. Hoornaert, *Newton polyhedra and Igusa's local zeta function*, J. Number Theory **89** (2001), no. 1, 31–64.

[13] J. Denef, F. Loeser, *Weights of exponential sums, intersection cohomology, and Newton polyhedra*, Invent. Math. **106** (1991), 275–294.

[14] ———, *Caractéristiques d'Euler-Poincaré, fonctions zêta locales et modifications analytiques*, J. Amer. Math. Soc. **5** (1992), no. 4, 705–720.

[15] ———, *Motivic Igusa Zeta functions*, J. Algebraic Geometry **7** (1998), 505–537.

[16] ———, *Germs of arcs on singular algebraic varieties and motivic integration*, Invent. Math. **135** (1999), 201–232.

[17] ———, *Geometry on arc spaces of algebraic varieties*, Proceedings of 3rd European Congress of Mathematics, Barcelona, Progress in Mathematics **201** (2001), Birkhäuser, 327-348.

[18] ———, *Motivic integration, quotient singularities and the McKay correspondence*, Compositio Math. **131** (2002), 267–290.

[19] Y.-N. Gau, *Embedded topological classification of quasi-ordinary singularities*, Mem. Amer. Math. Soc. **74** (1988), no. 388, 109–127.

[20] P.D. González Pérez, *Étude des singularités quasi-ordinaires d'hypersurfaces au moyen de la géométrie torique*, Thèse de Doctorat, Paris 7, 2002.

[21] ———, *The semigroup of a quasi-ordinary hypersurface*, Journal Inst. Math. Jussieu **2** (2003), 383–399.

[22] ———, *Toric embedded resolutions of quasi-ordinary hypersurface singularities*, Ann. Inst. Fourier **53** (2003), 1819–1881.

[23] P.D. González Pérez, L.J. McEwan, A. Némethi, *The zeta function of a quasi-ordinary singularity II*, Topics in algebraic and noncommutative geometry (Luminy/Annapolis, MD, 2001), 109–122, Contemp. Math., 324, Amer. Math. Soc., Providence, RI, 2003.

[24] G. Guibert, *Espaces d'arcs et invariants d'Alexander*, Comment. Math. Helv. **77** (2002), 783–820.

[25] J. Igusa, *Complex powers and asymptotic expansions I*, J. Reine Angew. Math. **268/269** (1974), 110–130.

[26] ———, *Complex powers and asymptotic expansions II*, J. Reine Angew. Math. **278/279** (1975), 307–321.

[27] S. Ishii, J. Kollár, *The Nash problem on arc families of singularities*, Duke Math. J. **120** (2003), no. 3, 601–620.

[28] H.W.E. Jung, *Darstellung der funktionen eines algebraischen Körpers zweier unabhaängigen Veränderlichen x, y in der Umgebung einer stelle $x = a, y = b$*, J. Reine Angew. Math. **133** (1908), 289–314.

[29] K. Hoornaert, D. Loots, *A computer program to calculate Igusa's Local Zeta Function and the Topological Zeta Function for non-degenerate polynomials*, freely available at http://www.wis.kuleuven.ac.be/algebra/kathleen/program.htm.

[30] H.B. Laufer, *Normal two-dimensional singularities*, Annals of Mathematics Studies **71**. Princeton University Press, Princeton, N.J.; University of Tokyo Press, Tokyo, 1971.

[31] J. Lipman, *Quasi-ordinary singularities of embedded surfaces*, Thesis, Harvard University (1965).

[32] ———, *Topological invariants of quasi-ordinary singularities*, Mem. Amer. Math. Soc. **74** (1988), no. 388, 1–107.

[33] F. Loeser, *Fonctions d'Igusa p-adiques et polynômes de Bernstein*, Amer. J. Math. **110** (1988), no. 1, 1–21.

[34] ———, *Fonctions d'Igusa p-adiques, polynômes de Bernstein, et polyèdres de Newton*, J. reine angew. Math. **412** (1990), 75–96.

[35] E. Looijenga, *Motivic measures.* Séminaire Bourbaki, Vol. 1999/2000. Astérisque, **276** (2002), 267–297.

[36] I. Luengo, *A new proof of the Jung-Abhyankar theorem*, J. Algebra **85** (1983), 399–409.

[37] L.J. McEwan, A. Némethi, *Some conjectures about quasi-ordinary singularities*, Topics in algebraic and noncommutative geometry (Luminy/Annapolis, MD, 2001), 185–193, Contemp. Math., **324**, Amer. Math. Soc., Providence, RI, 2003

[38] ———, *The zeta function of a quasi-ordinary singularity I*, Compositio Math. **140** (2004), no. 3, 667–682.

[39] P. Popescu-Pampu, *On the analytical invariance of the semigroups of a quasi-ordinary hypersurface singularity*, Duke Math. J. **124** (2004), no. 1, 67–104

[40] B. Rodrigues, W. Veys, *Holomorphy of Igusa's and topological zeta functions for homogeneous polynomials* Pacific J. Math. **201** (2001), no. 2, 429–440.

[41] J. Sebag, *Rationalité des séries de Poincaré et des fonctions zêta motiviques*, Manuscripta Math. **115** (2004), no. 2, 125–162.

[42] R. Stanley, *Enumerative combinatorics. Vol. I* With a foreword by Gian-Carlo Rota. The Wadsworth & Brooks/Cole Mathematics Series. Wadsworth & Brooks/Cole Advanced Books & Software, Monterey, CA, 1986.

[43] A Varchenko, *Zeta function of monodromy and Newton's diagram*, Invent. Math. **37** (1976), 253–262.

[44] W. Veys, *Poles of Igusa's local zeta function and monodromy*, Bull. Soc. Math. France **121** (1993), 545-598.

[45] ———, *Determination of the poles of the topological zeta function for curves*, Manuscripta Math. **87** (1995), 435–448.

[46] ———, *Zeta functions for curves and log canonical models*, Proc. London Math. Soc. **74** (1997), 360–378.

[47] ———, *Zeta functions and "Kontsevich invariants" on singular varieties*, Canad. J. Math. **53** (2001), no. 4, 834–865.

[48] O. Villamayor U., *On equiresolution and a question of Zariski*, Acta Math. **185** (2000), no. 1, 123–159.

[49] R.J. Walker, *Reduction of the singularities of an algebraic surface*, Annals Maths. **36** (1935), 336–365.

[50] O. Zariski, *Exceptional singularities of an algebroid surface and their reduction*, Atti. Accad. Naz. Lincei Rend., Cl. Sci. Fis. Mat. Natur., 8 43 (1967), 135–146. Reprinted in Collected papers, vol. 1 (1979).

[51] ———, *Contributions to the problem of equisingularity*, C.I.M.E. Questions on algebraic varieties. III ciclo, Varana 7–17 Settembre 1969, Edizione Cremonese, Roma 1970, 261–343. Reprinted in Collected papers, vol. 4 (1979).

Editorial Information

To be published in the *Memoirs*, a paper must be correct, new, nontrivial, and significant. Further, it must be well written and of interest to a substantial number of mathematicians. Piecemeal results, such as an inconclusive step toward an unproved major theorem or a minor variation on a known result, are in general not acceptable for publication. Papers appearing in *Memoirs* are generally at least 80 and not more than 200 published pages in length. Papers less than 80 or more than 200 published pages require the approval of the Managing Editor of the Transactions/Memoirs Editorial Board.

As of July 31, 2005, the backlog for this journal was approximately 14 volumes. This estimate is the result of dividing the number of manuscripts for this journal in the Providence office that have not yet gone to the printer on the above date by the average number of monographs per volume over the previous twelve months, reduced by the number of volumes published in four months (the time necessary for preparing a volume for the printer). (There are 6 volumes per year, each containing at least 4 numbers.)

A Consent to Publish and Copyright Agreement is required before a paper will be published in the *Memoirs*. After a paper is accepted for publication, the Providence office will send a Consent to Publish and Copyright Agreement to all authors of the paper. By submitting a paper to the *Memoirs*, authors certify that the results have not been submitted to nor are they under consideration for publication by another journal, conference proceedings, or similar publication.

Information for Authors

Memoirs are printed from camera copy fully prepared by the author. This means that the finished book will look exactly like the copy submitted.

The paper must contain a *descriptive title* and an *abstract* that summarizes the article in language suitable for workers in the general field (algebra, analysis, etc.). The *descriptive title* should be short, but informative; useless or vague phrases such as "some remarks about" or "concerning" should be avoided. The *abstract* should be at least one complete sentence, and at most 300 words. Included with the footnotes to the paper should be the 2000 *Mathematics Subject Classification* representing the primary and secondary subjects of the article. The classifications are accessible from `www.ams.org/msc/`. The list of classifications is also available in print starting with the 1999 annual index of *Mathematical Reviews*. The Mathematics Subject Classification footnote may be followed by a list of *key words and phrases* describing the subject matter of the article and taken from it. Journal abbreviations used in bibliographies are listed in the latest *Mathematical Reviews* annual index. The series abbreviations are also accessible from `www.ams.org/publications/`. To help in preparing and verifying references, the AMS offers MR Lookup, a Reference Tool for Linking, at `www.ams.org/mrlookup/`. When the manuscript is submitted, authors should supply the editor with electronic addresses if available. These will be printed after the postal address at the end of the article.

Electronically prepared manuscripts. The AMS encourages electronically prepared manuscripts, with a strong preference for AMS-LaTeX. To this end, the Society has prepared AMS-LaTeX author packages for each AMS publication. Author packages include instructions for preparing electronic manuscripts, the *AMS Author Handbook*, samples, and a style file that generates the particular design specifications of that publication series. Though AMS-LaTeX is the highly preferred format of TeX, author packages are also available in AMS-TeX.

Authors may retrieve an author package from e-MATH starting from `www.ams.org/tex/` or via FTP to `ftp.ams.org` (login as `anonymous`, enter username as password, and type `cd pub/author-info`). The *AMS Author Handbook* and the *Instruction Manual* are available in PDF format following the author packages link from `www.ams.org/tex/`. The author package can be obtained free of charge by sending email to pub@ams.org (Internet) or from the Publication Division, American Mathematical Society, 201 Charles St., Providence, RI 02904, USA. When requesting an author package, please specify $\mathcal{A}_{\mathcal{M}}\mathcal{S}$-LaTeX or $\mathcal{A}_{\mathcal{M}}\mathcal{S}$-TeX, Macintosh or IBM (3.5) format, and the publication in which your paper will appear. Please be sure to include your complete mailing address.

Sending electronic files. After acceptance, the source file(s) should be sent to the Providence office (this includes any TeX source file, any graphics files, and the DVI or PostScript file).

Before sending the source file, be sure you have proofread your paper carefully. The files you send must be the EXACT files used to generate the proof copy that was accepted for publication. For all publications, authors are required to send a printed copy of their paper, which exactly matches the copy approved for publication, along with any graphics that will appear in the paper.

TeX files may be submitted by email, FTP, or on diskette. The DVI file(s) and PostScript files should be submitted only by FTP or on diskette unless they are encoded properly to submit through email. (DVI files are binary and PostScript files tend to be very large.)

Electronically prepared manuscripts can be sent via email to pub-submit@ams.org (Internet). The subject line of the message should include the publication code to identify it as a Memoir. TeX source files, DVI files, and PostScript files can be transferred over the Internet by FTP to the Internet node e-math.ams.org (130.44.1.100).

Electronic graphics. Comprehensive instructions on preparing graphics are available at www.ams.org/jourhtml/graphics.html. A few of the major requirements are given here.

Submit files for graphics as EPS (Encapsulated PostScript) files. This includes graphics originated via a graphics application as well as scanned photographs or other computer-generated images. If this is not possible, TIFF files are acceptable as long as they can be opened in Adobe Photoshop or Illustrator. No matter what method was used to produce the graphic, it is necessary to provide a paper copy to the AMS.

Authors using graphics packages for the creation of electronic art should also avoid the use of any lines thinner than 0.5 points in width. Many graphics packages allow the user to specify a "hairline" for a very thin line. Hairlines often look acceptable when proofed on a typical laser printer. However, when produced on a high-resolution laser imagesetter, hairlines become nearly invisible and will be lost entirely in the final printing process.

Screens should be set to values between 15% and 85%. Screens which fall outside of this range are too light or too dark to print correctly. Variations of screens within a graphic should be no less than 10%.

Inquiries. Any inquiries concerning a paper that has been accepted for publication should be sent directly to the Electronic Prepress Department, American Mathematical Society, 201 Charles St., Providence, RI 02904, USA.

Editors

This journal is designed particularly for long research papers, normally at least 80 pages in length, and groups of cognate papers in pure and applied mathematics. Papers intended for publication in the *Memoirs* should be addressed to one of the following editors. In principle the Memoirs welcomes electronic submissions, and some of the editors, those whose names appear below with an asterisk (*), have indicated that they prefer them. However, editors reserve the right to request hard copies after papers have been submitted electronically. Authors are advised to make preliminary email inquiries to editors about whether they are likely to be able to handle submissions in a particular electronic form.

Titles in This Series

841 **Enrique Artal Bartolo, Pierrette Cassou-Noguès, Ignacio Luengo, and Alejandro Melle Hernández,** Quasi-ordinary power series and their zeta functions, 2005

840 **Sławomir Kołodziej,** The complex Monge-Ampère equation and pluripotential theory, 2005

839 **Mihai Ciucu,** A random tiling model for two dimensional electrostatics, 2005

838 **V. Jurdjevic,** Integrable Hamiltonian systems on complex Lie groups, 2005

837 **Joseph A. Ball and Victor Vinnikov,** Lax-Phillips scattering and conservative linear systems: A Cuntz-algebra multidimensional setting, 2005

836 **H. G. Dales and A. T.-M. Lau,** The second duals of Beurling algebras, 2005

835 **Kiyoshi Igusa,** Higher complex torsion and the framing principle, 2005

834 **Ken'ichi Ohshika,** Kleinian groups which are limits of geometrically finite groups, 2005

833 **Greg Hjorth and Alexander S. Kechris,** Rigidity theorems for actions of product groups and countable Borel equivalence relations, 2005

832 **Lee Klingler and Lawrence S. Levy,** Representation type of commutative Noetherian rings III: Global wildness and tameness, 2005

831 **K. R. Goodearl and F. Wehrung,** The complete dimension theory of partially ordered systems with equivalence and orthogonality, 2005

830 **Jason Fulman, Peter M. Neumann, and Cheryl E. Praeger,** A generating function approach to the enumeration of matrices in classical groups over finite fields, 2005

829 **S. G. Bobkov and B. Zegarlinski,** Entropy bounds and isoperimetry, 2005

828 **Joel Berman and Paweł M. Idziak,** Generative complexity in algebra, 2005

827 **Trevor A. Welsh,** Fermionic expressions for minimal model Virasoro characters, 2005

826 **Guy Métivier and Kevin Zumbrun,** Large viscous boundary layers for noncharacteristic nonlinear hyperbolic problems, 2005

825 **Yaozhong Hu,** Integral transformations and anticipative calculus for fractional Brownian motions, 2005

824 **Luen-Chau Li and Serge Parmentier,** On dynamical Poisson groupoids I, 2005

823 **Claus Mokler,** An analogue of a reductive algebraic monoid whose unit group is a Kac-Moody group, 2005

822 **Stefano Pigola, Marco Rigoli, and Alberto G. Setti,** Maximum principles on Riemannian manifolds and applications, 2005

821 **Nicole Bopp and Hubert Rubenthaler,** Local zeta functions attached to the minimal spherical series for a class of symmetric spaces, 2005

820 **Vadim A. Kaimanovich and Mikhail Lyubich,** Conformal and harmonic measures on laminations associated with rational maps, 2005

819 **F. Andreatta and E. Z. Goren,** Hilbert modular forms: Mod p and p-adic aspects, 2005

818 **Tom De Medts,** An algebraic structure for Moufang quadrangles, 2005

817 **Javier Fernández de Bobadilla,** Moduli spaces of polynomials in two variables, 2005

816 **Francis Clarke,** Necessary conditions in dynamic optimization, 2005

815 **Martin Bendersky and Donald M. Davis,** V_1-periodic homotopy groups of $SO(n)$, 2004

814 **Johannes Huebschmann,** Kähler spaces, nilpotent orbits, and singular reduction, 2004

813 **Jeff Groah and Blake Temple,** Shock-wave solutions of the Einstein equations with perfect fluid sources: Existence and consistency by a locally inertial Glimm scheme, 2004

812 **Richard D. Canary and Darryl McCullough,** Homotopy equivalences of 3-manifolds and deformation theory of Kleinian groups, 2004

811 **Ottmar Loos and Erhard Neher,** Locally finite root systems, 2004

810 **W. N. Everitt and L. Markus,** Infinite dimensional complex symplectic spaces, 2004

809 **J. T. Cox, D. A. Dawson, and A. Greven,** Mutually catalytic super branching random walks: Large finite systems and renormalization analysis, 2004

808 **Hagen Meltzer,** Exceptional vector bundles, tilting sheaves and tilting complexes for weighted projective lines, 2004

807 **Carlos A. Cabrelli, Christopher Heil, and Ursula M. Molter,** Self-similarity and multiwavelets in higher dimensions, 2004

806 **Spiros A. Argyros and Andreas Tolias,** Methods in the theory of hereditarily indecomposable Banach spaces, 2004

805 **Philip L. Bowers and Kenneth Stephenson,** Uniformizing dessins and Belyĭ maps via circle packing, 2004

804 **A. Yu Ol'shanskii and M. V. Sapir,** The conjugacy problem and Higman embeddings, 2004

803 **Michael Field and Matthew Nicol,** Ergodic theory of equivariant diffeomorphisms: Markov partitions and stable ergodicity, 2004

802 **Martin W. Liebeck and Gary M. Seitz,** The maximal subgroups of positive dimension in exceptional algebraic groups, 2004

801 **Fabio Ancona and Andrea Marson,** Well-posedness for general 2×2 systems of conservation law, 2004

800 **V. Poénaru and C. Tanas,** Equivariant, almost-arborescent representation of open simply-connected 3-manifolds; A finiteness result, 2004

799 **Barry Mazur and Karl Rubin,** Kolyvagin systems, 2004

798 **Benoît Mselati,** Classification and probabilistic representation of the positive solutions of a semilinear elliptic equation, 2004

797 **Ola Bratteli, Palle E. T. Jorgensen, and Vasyl' Ostrovs'kyĭ,** Representation theory and numerical AF-invariants, 2004

796 **Marc A. Rieffel,** Gromov-Hausdorff distance for quantum metric spaces/Matrix algebras converge to the sphere for quantum Gromov-Hausdorff distance, 2004

795 **Adam Nyman,** Points on quantum projectivizations, 2004

794 **Kevin K. Ferland and L. Gaunce Lewis, Jr.,** The $RO(G)$-graded equivariant ordinary homology of G-cell complexes with even-dimensional cells for $G = \mathbb{Z}/p$, 2004

793 **Jindřich Zapletal,** Descriptive set theory and definable forcing, 2004

792 **Inmaculada Baldomá and Ernest Fontich,** Exponentially small splitting of invariant manifolds of parabolic points, 2004

791 **Eva A. Gallardo-Gutiérrez and Alfonso Montes-Rodríguez,** The role of the spectrum in the cyclic behavior of composition operators, 2004

790 **Thierry Lévy,** Yang-Mills measure on compact surfaces, 2003

789 **Helge Glöckner,** Positive definite functions on infinite-dimensional convex cones, 2003

788 **Robert Denk, Matthias Hieber, and Jan Prüss,** $\mathcal{R}$-boundedness, Fourier multipliers and problems of elliptic and parabolic type, 2003

787 **Michael Cwikel, Per G. Nilsson, and Gideon Schechtman,** Interpolation of weighted Banach lattices/A characterization of relatively decomposable Banach lattices, 2003

786 **Arnd Scheel,** Radially symmetric patterns of reaction-diffusion systems, 2003

785 **R. R. Bruner and J. P. C. Greenlees,** The connective K-theory of finite groups, 2003

784 **Desmond Sheiham,** Invariants of boundary link cobordism, 2003

For a complete list of titles in this series, visit the AMS Bookstore at **www.ams.org/bookstore/**.